Classical Electrodynamics

Classical Electrodynamics: A Concise and Detailed Guide covers the essential theoretical foundations of electrodynamics from the vector calculus formulation to the classical gauge field theory.

The essential theoretical formalism of electrodynamics is covered with all mathematical derivations provided. The derivations are uniquely presented with inline, line-by-line explanations, which makes every derivation easy to follow. The reader does not need to struggle to fill in the steps, and the reader can focus on pondering over the physical content and implications of the derived results. The content is supplemented with numerous exercises with their complete solutions included.

This book is for students who are interested in theoretical physics and seek an efficient but proper preparation of the essential knowledge in electrodynamics and classical gauge field theory. With this background knowledge, the student will be able to progress further into concepts beyond electrodynamics, like general relativity or quantum gauge field theories.

Meng Lee Leek is a Senior Lecturer of Physics at Nanyang Technological University of Singapore (NTU). He has been awarded the Nanyang Education Award once and the Teaching Excellence Award 3 times. His research interests lie in theoretical physics, and he enjoys showing students the world of theoretical physics through supervision of numerous student projects.

Classical Electrodynamics
A Concise and Detailed Guide

Meng Lee Leek

CRC Press is an imprint of the
Taylor & Francis Group, an **informa** business

Designed cover image: Meng Lee Leek, CRC Press

First edition published 2025
by CRC Press
2385 NW Executive Center Drive, Suite 320, Boca Raton FL 33431

and by CRC Press
4 Park Square, Milton Park, Abingdon, Oxon, OX14 4RN

CRC Press is an imprint of Taylor & Francis Group, LLC

© 2025 Meng Lee Leek

Reasonable efforts have been made to publish reliable data and information, but the author and publisher cannot assume responsibility for the validity of all materials or the consequences of their use. The authors and publishers have attempted to trace the copyright holders of all material reproduced in this publication and apologize to copyright holders if permission to publish in this form has not been obtained. If any copyright material has not been acknowledged please write and let us know so we may rectify in any future reprint.

Except as permitted under U.S. Copyright Law, no part of this book may be reprinted, reproduced, transmitted, or utilized in any form by any electronic, mechanical, or other means, now known or hereafter invented, including photocopying, microfilming, and recording, or in any information storage or retrieval system, without written permission from the publishers.

For permission to photocopy or use material electronically from this work, access www.copyright.com or contact the Copyright Clearance Center, Inc. (CCC), 222 Rosewood Drive, Danvers, MA 01923, 978-750-8400. For works that are not available on CCC please contact mpkbookspermissions@tandf.co.uk

Trademark notice: Product or corporate names may be trademarks or registered trademarks and are used only for identification and explanation without intent to infringe.

ISBN: 978-1-032-84831-0 (hbk)
ISBN: 978-1-032-84832-7 (pbk)
ISBN: 978-1-003-51521-0 (ebk)

DOI: 10.1201/9781003515210

Typeset in Latin Modern font
by KnowledgeWorks Global Ltd.

Publisher's note: This book has been prepared from camera-ready copy provided by the authors.

Contents

Preface   ix

I   Basic Formalism of Electrodynamics (In vector calculus form)   1

1   Conservation Laws   3
- 1.1   Charge continuity equation   3
- 1.2   Energy continuity equation   4
- 1.3   Momentum continuity equation   5
- 1.4   Angular momentum   8

2   Potential Formulation of Electrodynamics   9
- 2.1   Electrodynamics in terms of potentials   9
- 2.2   Gauge transformations   10
  - 2.2.1   Coulomb gauge   11
  - 2.2.2   Lorenz gauge   12
- 2.3   Solutions are retarded potentials   12
- 2.4   The fields: Jefimenko's equations   19

3   Standard Application 1: Point Charge   21
- 3.1   Retarded potentials: Lienard–Wiechert potentials   21
- 3.2   Fields   24
- 3.3   Example: Point charge in constant velocity   26

4   Standard Application 2: Radiation   30
- 4.1   From point charge   30
- 4.2   From hertzian dipole   32
- 4.3   From arbitrary distribution   36

5   Standard Application 3: Scattering of EM waves   39
- 5.1   Statement of the scattering problem   39
- 5.2   Formalism: Optical theorem   40
- 5.3   Basic scattering models   44
  - 5.3.1   Scattering by free charge: Thomson scattering   44
  - 5.3.2   Scattering by bound charge   46

6   Exercises for Part I   49
- 6.1   Maxwell stress tensor   49
- 6.2   Existence of the gauges   49
- 6.3   Check back the retarded potential solutions   49
- 6.4   Retarded potentials satisfy Lorenz gauge   50
- 6.5   Quasistatic approximation of Jefimenko's equation   50
- 6.6   Charge moving at constant velocity   50
- 6.7   Decay of the classical atom   51

	6.8	Dipole retarded potentials satisfy Lorenz gauge	51
	6.9	Recover various cases	51
	6.10	Magnetic dipole radiation	51
	6.11	Thomson scattering is nonrelativistic	52
	6.12	Satisfies optical theorem	52
	6.13	Dielectric sphere scatterer	52

II Electrodynamics in Lorentz Covariant Form (in tensor calculus form) — 53

7 Special Relativity — 55

- 7.1 Derivation of Lorentz transformation — 55
- 7.2 Basic consequences of special relativity — 59
  - 7.2.1 Time dilation — 59
  - 7.2.2 Length contraction — 59
  - 7.2.3 Velocity addition — 60
- 7.3 Further consequences (Minkowski spacetime diagram, invariant interval and causality) — 61
  - 7.3.1 Minkowski spacetime diagram — 61
  - 7.3.2 Invariant interval — 61
- 7.4 Transition into tensor calculus — 63
  - 7.4.1 Minkowski metric, 4-vectors — 63
  - 7.4.2 Momentum 4-vector and Einstein relation — 65
  - 7.4.3 Derivative 4-vectors and d'Alembert's operator — 67
  - 7.4.4 Relativistic mechanics — 69

8 Electrodynamics Recast into a Manifestly Lorentz Covariant Form — 71

- 8.1 Just for motivation: Magnetism as a relativistic phenomena — 71
- 8.2 Transformations of $\vec{E}$ and $\vec{B}$ — 73
- 8.3 Rewriting electrodynamics or unification of $\vec{E}$ and $\vec{B}$ — 76
  - 8.3.1 Maxwell's equations — 76
  - 8.3.2 Conservation laws — 81
  - 8.3.3 Lorentz invariants in electrodynamics — 83
- 8.4 Point charge revisited: Lienard–Wiechert potentials — 84
- 8.5 Point charge revisited: Fields of a charged particle in constant velocity — 86
- 8.6 Point charge revisited: Lienard's generalisation of Larmor's formula — 88

9 Exercises for Part II — 94

- 9.1 Poincare group — 94
- 9.2 Minkowski diagram exercises — 94
- 9.3 Photon — 95
- 9.4 Checking Lorentz transformations — 95
- 9.5 Maxwell's equations — 95
- 9.6 Meaning of K^0 — 95
- 9.7 Gauge invariance — 96
- 9.8 Lorentz invariants — 96
- 9.9 Lienard–Wiechert potentials in covariant form — 96
- 9.10 Radiation for acceleration parallel to velocity — 96

III Classical Relativistic $U(1)$ Gauge Theory — 97

10 Lagrangian Description of a Classical Relativistic $U(1)$ Gauge Theory — 99
- 10.1 Review of Lagrangian mechanics and Noether's theorem — 99
- 10.2 Relativistic Lagrangian of a charged particle interacting with external EM field — 100
- 10.3 Lagrangian of the EM field — 104
  - 10.3.1 Where to go after this book — 111

11 Exercises for Part III — 112
- 11.1 Checking expressions — 112

IV Solutions to the Problems — 113

12 Solutions to Exercises in Part I — 115
- 12.1 Solution to exercise 6.1 — 115
- 12.2 Solution to exercise 6.2 — 116
- 12.3 Solution to exercise 6.3 — 117
- 12.4 Solution to exercise 6.4 — 118
- 12.5 Solution to exercise 6.5 — 119
- 12.6 Solution to exercise 6.6 — 120
- 12.7 Solution to exercise 6.7 — 122
- 12.8 Solution to exercise 6.8 — 123
- 12.9 Solution to exercise 6.9 — 124
- 12.10 Solution to exercise 6.10 — 125
- 12.11 Solution to exercise 6.11 — 125
- 12.12 Solution to exercise 6.12 — 125
- 12.13 Solution to exercise 6.13 — 127

13 Solutions to Exercises for Part II — 129
- 13.1 Solution to exercise 9.1 — 129
- 13.2 Solution to exercise 9.2 — 129
- 13.3 Solution to exercise 9.3 — 132
- 13.4 Solution to exercise 9.4 — 132
- 13.5 Solution to exercise 9.5 — 132
- 13.6 Solution to exercise 9.6 — 133
- 13.7 Solution to exercise 9.7 — 133
- 13.8 Solution to exercise 9.8 — 133
- 13.9 Solution to exercise 9.9 — 135
- 13.10 Solution to exercise 9.10 — 135

14 Solutions to Exercises for Part III — 137
- 14.1 Solution to exercise 11.1 — 137

V Appendices — 141

15 Mathematics: Vector Calculus — 143
- 15.1 3 Coordinate systems — 143
  - 15.1.1 Cartesian coordinates — 143
  - 15.1.2 Spherical coordinates — 143
  - 15.1.3 Cylindrical coordinates — 144

	15.2	Vector identities	144
		15.2.1 Triple products	144
		15.2.2 Product rules	144
		15.2.3 Second derivatives	145
		15.2.4 Integral theorems	145
		15.2.5 Dirac delta function	145

16 Summary of Electromagnetism — 146

	16.1	Electrostatics	146
	16.2	Electric fields in matter	147
	16.3	Magnetostatics	148
	16.4	Magnetic fields in matter	149
	16.5	Further topics	150
	16.6	Electromagnetic waves (Standard application of electrodynamics)	151
	16.7	Exercises	152
		16.7.1 Evaluating line integral	152
		16.7.2 Proving identity	152
		16.7.3 Proving identity	153
		16.7.4 Electrostatics	153
		16.7.5 Electric dipole	153
		16.7.6 Magnetostatics	153
		16.7.7 Magnetic field of a surface current	153
		16.7.8 Poynting vector	153
		16.7.9 Electromagnetic waves	154
	16.8	Solutions to Exercises	154
		16.8.1 Solution to exercise 16.7.1	154
		16.8.2 Solution to exercise 16.7.2	154
		16.8.3 Solution to exercise 16.7.3	155
		16.8.4 Solution to exercise 16.7.4	155
		16.8.5 Solution to exercise 16.7.5	156
		16.8.6 Solution to exercise 16.7.6	156
		16.8.7 Solution to exercise 16.7.7	157
		16.8.8 Solution to exercise 16.7.8	157
		16.8.9 Solution to exercise 16.7.9	158

Bibliography — 159

Index — 161

Preface

This book grew out of my lectures notes used in teaching the course PH4506 Electrodynamics at Nanyang Technological University (NTU), School of Physical and Mathematical Sciences (SPMS).

By tradition, electrodynamics is taught in year 4 but personally I feel that it is too late as students are working on their final year projects (FYPs) at the same time where electrodynamics is clearly an essential foundational topic. This is especially true if their FYPs are related to quantum field theory.

Thus I designed these notes to elevate students to reach the most advanced theoretical description of electrodynamics which is in the form of a classical field theory with a $U(1)$ gauge group in terms of tensor calculus. This is also one of the purposes of publishing my notes into this book: to provide a concise journey to the most advanced theoretical description of electrodynamics.

In addition, I have also provided detailed steps throughout so that students can follow the calculations with ease and this they can divert their time to try to understand the physical content of the calculations rather than busily trying to fill in the missing steps. This feature was quite highly appreciated by my students as they felt that they have the "spare time", given the tight duration of a 3-month semester, to think about the purpose and implications of the calculations done in the lecture notes.

I hope undergraduate students can start reading this book right after your year-2 Electromagnetism course. This book will then serve as an introduction to tensor calculus and elevate your knowledge of electrodynamics to the level of a classical abelian gauge theory. This will prepare you to learn quantum field theory before your FYP.

Finally, I would like to declare that all the content here in this book is not original. My writing has been inspired by many of the popular and classic textbooks on electrodynamics for instance [4], [6], [5], [3] and [7].

Part I

Basic Formalism of Electrodynamics
(In vector calculus form)

1

Conservation Laws

Conservation laws are important in a theory because they express the conserved quantities in the theory. More deeply, they express the symmetries inherent in the theory (via Noether's theorem).

1.1 Charge continuity equation

The global conservation of electric charge means that the total charge in the universe is a constant. The stronger statement is the local conservation of charge, which states: if the total charge in some volume changes, then that amount of charge must have entered or exited through the closed surface bounding that volume. We want to obtain the mathematical expression for that statement.[1]

$$\frac{dQ}{dt} = -\oint \vec{J} \cdot d\vec{A} \tag{1.1}$$

$\qquad$ | where $\dfrac{dQ}{dt}$ means rate of change of charge in the volume

$\qquad$ | where $\oint \vec{J} \cdot d\vec{A}$ means net flux of current over the bounding surface

$\qquad$ | where the negative sign indicates a decrease in charge Q

$\qquad$ | write $Q(t) = \int \rho(\vec{r}, t) dV$

$$\int \frac{\partial \rho(\vec{r}, t)}{\partial t} dV = -\oint \vec{J} \cdot d\vec{A} \tag{1.2}$$

$\qquad$ | use divergence theorem

$$\int \frac{\partial \rho(\vec{r}, t)}{\partial t} dV = -\int \vec{\nabla} \cdot \vec{J} dV \tag{1.3}$$

$\qquad$ | since it holds true for any volume

$$\boxed{\frac{\partial \rho(\vec{r}, t)}{\partial t} = -\vec{\nabla} \cdot \vec{J}(\vec{r}, t)} \tag{1.4}$$

[2] Note that we may obtain the same continuity equation from Maxwell's equations, but the displacement current term was added to be compatible with the charge continuity equation, so the argument may be somewhat circular.

---

[1] This is really an assumption and we are just writing a mathematical statement for the assumption.
[2] Note the general pattern of the continuity equation: $\frac{\partial \text{"density"}}{\partial t} = -\vec{\nabla} \cdot \text{"current"}$.

1.2 Energy continuity equation

Conservation of energy is of course one of the most sacred conservation laws we believe in. Actually in the context of relativity, conservation of energy[3] and conservation of momentum are one combined conservation law, as we will see in section 8.3.2. Now we will handle the laws of conservation of energy and conservation of momentum separately.

Consider work done by electric and magnetic forces on charge q,

$$dW = \vec{F} \cdot d\vec{l} = q(\vec{E} + \vec{v} \times \vec{B}) \cdot \vec{v} dt \tag{1.5}$$

| due to the dot product, magnetic forces do no work

$$dW = q\vec{E} \cdot \vec{v} dt \tag{1.6}$$

$$dW = \rho dV \vec{E} \cdot \vec{v} dt \tag{1.7}$$

| recall that $\rho \vec{v} = \vec{J}$

$$\frac{dW}{dt} = \int \vec{E} \cdot \vec{J} dV \tag{1.8}$$

Thus $\vec{E} \cdot \vec{J}$ is the power (per unit volume) delivered to the charges.

$$\frac{dW}{dt} = \int \vec{E} \cdot \vec{J} dV \tag{1.9}$$

$$\frac{dW}{dt} = \int \vec{E} \cdot \left(\frac{1}{\mu_0} \vec{\nabla} \times \vec{B} - \epsilon_0 \frac{\partial \vec{E}}{\partial t} \right) dV \tag{1.10}$$

| where we replace $\vec{J}$ using Ampere–Maxwell law

$$\frac{dW}{dt} = \int \left(\frac{1}{\mu_0} \vec{E} \cdot (\vec{\nabla} \times \vec{B}) - \epsilon_0 \vec{E} \cdot \frac{\partial \vec{E}}{\partial t} \right) dV \tag{1.11}$$

| use identity $\vec{\nabla} \cdot (\vec{E} \times \vec{B}) = \vec{B} \cdot (\vec{\nabla} \times \vec{E}) - \vec{E} \cdot (\vec{\nabla} \times \vec{B})$

| then use Faraday's law, $\vec{E} \cdot (\vec{\nabla} \times \vec{B}) = -\vec{\nabla} \cdot (\vec{E} \times \vec{B}) + \vec{B} \cdot \left(-\frac{\partial \vec{B}}{\partial t} \right)$

| then rewrite $\vec{B} \cdot \frac{\partial \vec{B}}{\partial t} = \frac{1}{2} \frac{\partial}{\partial t} \vec{B}^2$ and $\vec{E} \cdot \frac{\partial \vec{E}}{\partial t} = \frac{1}{2} \frac{\partial}{\partial t} \vec{E}^2$

$$\frac{dW}{dt} = -\frac{d}{dt} \int \frac{1}{2} \left(\epsilon_0 \vec{E}^2 + \frac{1}{\mu_0} \vec{B}^2 \right) dV - \frac{1}{\mu_0} \oint (\vec{E} \times \vec{B}) \cdot d\vec{A} \tag{1.12}$$

| define the EM field energy density, $u_{\text{EM}} = \frac{1}{2} \left(\epsilon_0 \vec{E}^2 + \frac{1}{\mu_0} \vec{B}^2 \right)$

| and define the Poynting vector, $\vec{S} = \frac{1}{\mu_0} (\vec{E} \times \vec{B})$

$$\frac{dW}{dt} = -\frac{d}{dt} \int u_{\text{EM}} dV - \oint \vec{S} \cdot d\vec{A} \tag{1.13}$$

where the closed surface bounds volume V. This is Poynting's theorem or "work-energy theorem" of electrodynamics. We will derive the differential version or the energy continuity equation shortly.

---

[3] Which includes rest mass energy.

Conservation Laws

We first need to interpret Poynting's theorem clearly:

- $\frac{dW}{dt}$ is the work done (per unit time) on the charges in volume V by EM forces
- $-\frac{d}{dt} \int u_{\text{EM}} dV$ is the change in energy (per unit time) stored in the fields
- $-\oint \vec{S} \cdot d\vec{A}$ is the energy (per unit time) that flowed out through the surface

Suppose we can write a mechanical energy density u_{mech}, then $\frac{dW}{dt} = \frac{d}{dt} \int u_{\text{mech}} dV$, and a differential version can be written:

$$\frac{d}{dt} \int u_{\text{mech}} dV = -\frac{d}{dt} \int u_{\text{EM}} dV - \oint \vec{S} \cdot d\vec{A} \qquad (1.14)$$

| use divergence theorem on the last term
| since the equation holds for any volume

$$\boxed{\frac{\partial}{\partial t}(u_{\text{mech}} + u_{\text{EM}}) = -\vec{\nabla} \cdot \vec{S}} \qquad (1.15)$$

which is the continuity equation for energy and the Poynting vector is the energy current vector. If there are no sources in volume V, we have $\dfrac{\partial u_{\text{EM}}}{\partial t} = -\vec{\nabla} \cdot \vec{S}$, which simply means that the change in energy stored in the field in volume V is equal to the amount of energy that flowed in or out of the surface bounding volume V.

1.3 Momentum continuity equation

Let us look at a quick example to see why the EM field also carries momentum.

Consider two charges moving with constant v, as shown in figure 1.1.

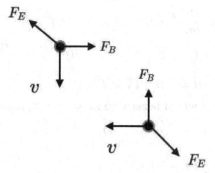

FIGURE 1.1
Two charges moving with speed v and exerting forces on each other.

The calculation of E-field and B-field due to moving charges is not from Coulomb's law and Biot-Savart law.[4] This calculation will be done later in section 3.3, but the essential

---
[4]Both laws are only true in the low-speed limit of electrodynamics.

results are easily stated: the E-field is still radial, and the B-field is still circular. Thus the electric force and magnetic force are as drawn in figure 1.1, and we can see that the net force on each charge is equal in magnitude but not opposite to each other.

This appears to be a violation of Newton's third law, which is essential in the conservation of momentum. Thus if we realise that the EM fields also carry momentum, then the conservation of momentum would be the conservation of field momentum + mechanical momentum.

We shall now derive the momentum continuity equation. Consider the total electromagnetic force on charges in volume V,

$$\vec{F} = \int \rho(\vec{E} + \vec{v} \times \vec{B})dV \tag{1.16}$$

| recall that $\rho\vec{v} = \vec{J}$

$$\vec{F} = \int (\rho\vec{E} + \vec{J} \times \vec{B})dV \tag{1.17}$$

We can denote $\vec{f} = \rho\vec{E} + \vec{J} \times \vec{B}$ as the force per unit volume. We want to eliminate ρ and $\vec{J}$ and express the RHS completely in terms of the fields. This will allow us to discover the field momentum (and the Maxwell Stress tensor).

$$\vec{f} = \rho\vec{E} + \vec{J} \times \vec{B} \tag{1.18}$$

| use Gauss law: $\rho = \epsilon_0(\vec{\nabla} \cdot \vec{E})$

| and Ampere–Maxwell law: $\vec{J} = \dfrac{1}{\mu_0}\vec{\nabla} \times \vec{B} - \epsilon_0 \dfrac{\partial \vec{E}}{\partial t}$

$$\vec{f} = \epsilon_0(\vec{\nabla} \cdot \vec{E})\vec{E} + \left(\dfrac{1}{\mu_0}\vec{\nabla} \times \vec{B} - \epsilon_0 \dfrac{\partial \vec{E}}{\partial t}\right) \times \vec{B} \tag{1.19}$$

| then use $\dfrac{\partial}{\partial t}(\vec{E} \times \vec{B}) = \dfrac{\partial \vec{E}}{\partial t} \times \vec{B} + \vec{E} \times \dfrac{\partial \vec{B}}{\partial t}$

| and Faraday's law: $\dfrac{\partial \vec{B}}{\partial t} = -\vec{\nabla} \times \vec{E}$

| so that $\dfrac{\partial \vec{E}}{\partial t} \times \vec{B} = \dfrac{\partial}{\partial t}(\vec{E} \times \vec{B}) + \vec{E} \times (\vec{\nabla} \times \vec{E})$

$$\vec{f} = \epsilon_0\left[(\vec{\nabla} \cdot \vec{E})\vec{E} - \vec{E} \times (\vec{\nabla} \times \vec{E})\right] + \dfrac{1}{\mu_0}(\vec{\nabla} \times \vec{B}) \times \vec{B} - \epsilon_0\dfrac{\partial}{\partial t}(\vec{E} \times \vec{B}) \tag{1.20}$$

| rewrite $(\vec{\nabla} \times \vec{B}) \times \vec{B} = -\vec{B} \times (\vec{\nabla} \times \vec{B})$

| introduce $(\vec{\nabla} \cdot \vec{B})\vec{B}$ which is zero, to make the expression more symmetrical

| recall the Poynting vector: $\vec{S} = \dfrac{1}{\mu_0}(\vec{E} \times \vec{B})$

$$\vec{f} = \epsilon_0\left[(\vec{\nabla} \cdot \vec{E})\vec{E} - \vec{E} \times (\vec{\nabla} \times \vec{E})\right]$$
$$+ \dfrac{1}{\mu_0}\left[(\vec{\nabla} \cdot \vec{B})\vec{B} - \vec{B} \times (\vec{\nabla} \times \vec{B})\right] - \epsilon_0\mu_0\dfrac{\partial \vec{S}}{\partial t} \tag{1.21}$$

| use identity: $\vec{\nabla}\vec{E}^2 = 2(\vec{E} \cdot \vec{\nabla})\vec{E} + 2\vec{E} \times (\vec{\nabla} \times \vec{E})$

| $\Longrightarrow \vec{E} \times (\vec{\nabla} \times \vec{E}) = \dfrac{1}{2}\vec{\nabla}\vec{E}^2 - (\vec{E} \cdot \vec{\nabla})\vec{E}$

| a similar identity holds for $\vec{B}$: $\vec{B} \times (\vec{\nabla} \times \vec{B}) = \dfrac{1}{2}\vec{\nabla}\vec{B}^2 - (\vec{B} \cdot \vec{\nabla})\vec{B}$

$$\begin{aligned}\vec{f} &= \epsilon_0\left[(\vec{\nabla}\cdot\vec{E})\vec{E}+(\vec{E}\cdot\vec{\nabla})\vec{E}\right]+\frac{1}{\mu_0}\left[(\vec{\nabla}\cdot\vec{B})\vec{B}+(\vec{B}\cdot\vec{\nabla})\vec{B}\right]\\&\quad-\frac{1}{2}\vec{\nabla}\left(\epsilon_0\vec{E}^2+\frac{1}{\mu_0}\vec{B}^2\right)-\epsilon_0\mu_0\frac{\partial\vec{S}}{\partial t}\end{aligned}\tag{1.22}$$

| introduce the 3×3 matrix: Maxwell stress tensor
| which is: $T_{ij}=\epsilon_0\left(E_iE_j-\frac{1}{2}\delta_{ij}\vec{E}^2\right)+\frac{1}{\mu_0}\left(B_iB_j-\frac{1}{2}\delta_{ij}\vec{B}^2\right)$
| the divergence of $\overleftrightarrow{T}$ is $(\vec{\nabla}\cdot\overleftrightarrow{T})_j=\sum_{i=x,y,z}\nabla_iT_{ij}$
| and note that $\sum_{i=x,y,z}\nabla_iE_i=\vec{\nabla}\cdot\vec{E}$ and $\sum_{i=x,y,z}E_i\nabla_iE_j=(\vec{E}\cdot\vec{\nabla})E_j$
| we have $(\vec{\nabla}\cdot\overleftrightarrow{T})_j=\epsilon_0\left((\vec{\nabla}\cdot\vec{E})E_j+(\vec{E}\cdot\vec{\nabla})E_j-\frac{1}{2}\nabla_j\vec{E}^2\right)$
| $\qquad\qquad\qquad+\frac{1}{\mu_0}\left((\vec{\nabla}\cdot\vec{B})B_j+(\vec{B}\cdot\vec{\nabla})B_j-\frac{1}{2}\nabla_j\vec{B}^2\right)$

$$\vec{f}=\vec{\nabla}\cdot\overleftrightarrow{T}-\epsilon_0\mu_0\frac{\partial\vec{S}}{\partial t}\tag{1.23}$$

The total force is obtained by integrating over volume V.

$$\vec{F}=\int\vec{\nabla}\cdot\overleftrightarrow{T}\,dV-\epsilon_0\mu_0\int\frac{\partial\vec{S}}{\partial t}dV\tag{1.24}$$

| use divergence theorem

$$\vec{F}=\oint\overleftrightarrow{T}\cdot d\vec{A}-\epsilon_0\mu_0\frac{d}{dt}\int\vec{S}dV\tag{1.25}$$

Thus we can interpret $\overleftrightarrow{T}$ as a generalised pressure. The diagonal elements, T_{xx}, T_{yy} and T_{zz}, are pressures and the off-diagonal elements, T_{xy}, T_{yz}, ... are shears.[5]

We want to write it into the differential version, which is the momentum continuity equation.

$$\int\vec{\nabla}\cdot\overleftrightarrow{T}\,dV-\epsilon_0\mu_0\frac{d}{dt}\int\vec{S}dV=\vec{F}\tag{1.26}$$

define a mechanical momentum density $\vec{F}=\frac{d}{dt}\int\vec{p}_{\text{mech}}dV$ |

the field momentum density is $\vec{p}_{\text{EM}}=\epsilon_0\mu_0\vec{S}=\frac{1}{c^2}\vec{S}$ |

$$\frac{d}{dt}\int(\vec{p}_{\text{mech}}+\vec{p}_{\text{EM}})dV=\int\vec{\nabla}\cdot\overleftrightarrow{T}\,dV\tag{1.27}$$

$$\boxed{\frac{\partial}{\partial t}(\vec{p}_{\text{mech}}+\vec{p}_{\text{EM}})=\vec{\nabla}\cdot\overleftrightarrow{T}}\tag{1.28}$$

which is the momentum continuity equation. We can thus also interpret $-\overleftrightarrow{T}$ as the momentum flux density.

---

[5] Because diagonal elements mean "same direction", for example, x affecting x; but off-diagonal means "different direction", for example, T_{xy} means "applied in x direction but affects y direction".

1.4 Angular momentum

For field angular momentum, we shall merely state the definition of the angular momentum (density) to be the usual

$$l_{\text{EM}} = \vec{r} \times \vec{p}_{\text{EM}} = \epsilon_0 \left[\vec{r} \times (\vec{E} \times \vec{B}) \right] \tag{1.29}$$

For more details, please refer to the classic textbook by [5].

2

Potential Formulation of Electrodynamics

2.1 Electrodynamics in terms of potentials

[1] In electrostatics and magnetostatics, it is advantageous to solve the potential problem (by solving the Poisson or Laplace equation) then get the fields from the potentials. In electrodynamics, the advantage is smaller. So it is still worthwhile to express electrodynamics in terms of potentials. This means we want the time dependent solutions of ϕ and $\vec{A}$ resulting from time dependent sources $\rho(\vec{r}, t)$ and $\vec{J}(\vec{r}, t)$.

Since $\vec{B}$ is still divergenceless (because there are no magnetic monopoles) in the time dependent problem, we still have

$$\vec{B} = \vec{\nabla} \times \vec{A} \tag{2.1}$$

Curl of $\vec{E}$ is Faraday's law, so

$$\vec{\nabla} \times \vec{E} = -\frac{\partial \vec{B}}{\partial t} \tag{2.2}$$

$$\vert \text{ insert } \vec{B} = \vec{\nabla} \times \vec{A} \text{ and move it to LHS}$$

$$\vec{\nabla} \times \left(\vec{E} + \frac{\partial \vec{A}}{\partial t} \right) = 0 \tag{2.3}$$

$$\vert \text{ recall the identity (section 15.2.3): } \vec{\nabla} \times (\vec{\nabla} f) = 0$$

$$\vec{E} + \frac{\partial \vec{A}}{\partial t} = -\vec{\nabla} \phi \tag{2.4}$$

$$\vec{E} = -\vec{\nabla} \phi - \frac{\partial \vec{A}}{\partial t} \tag{2.5}$$

When we insert this $\vec{E}$ into Gauss law,

$$\vec{\nabla} \cdot \vec{E} = \frac{\rho}{\epsilon_0} \tag{2.6}$$

$$\vec{\nabla} \cdot \left(-\vec{\nabla}\phi - \frac{\partial \vec{A}}{\partial t} \right) = \frac{\rho}{\epsilon_0} \tag{2.7}$$

$$\boxed{\vec{\nabla}^2 \phi + \frac{\partial}{\partial t}(\vec{\nabla} \cdot \vec{A}) = -\frac{\rho}{\epsilon_0}} \tag{2.8}$$

which is the time-dependent generalisation of Poisson equation.

---

[1] The potential method is useful as we usually know the sources and then we can solve for the potentials.

Now using Ampere–Maxwell equation

$$\vec{\nabla} \times \vec{B} = \mu_0 \vec{J} + \epsilon_0 \mu_0 \frac{\partial \vec{E}}{\partial t} \tag{2.9}$$

$$| \text{ insert } \vec{B} = \vec{\nabla} \times \vec{A} \text{ and } \vec{E} = -\vec{\nabla}\phi - \frac{\partial \vec{A}}{\partial t}$$

$$\vec{\nabla} \times (\vec{\nabla} \times \vec{A}) = \mu_0 \vec{J} - \mu_0 \epsilon_0 \vec{\nabla}\left(\frac{\partial \phi}{\partial t}\right) - \mu_0 \epsilon_0 \frac{\partial^2 \vec{A}}{\partial t^2} \tag{2.10}$$

$$| \text{ use vector identity } \vec{\nabla} \times (\vec{\nabla} \times \vec{A}) = \vec{\nabla}(\vec{\nabla} \cdot \vec{A}) - \vec{\nabla}^2 \vec{A}$$

$$\boxed{-\mu_0 \vec{J} = \left(\vec{\nabla}^2 \vec{A} - \mu_0 \epsilon_0 \frac{\partial^2 \vec{A}}{\partial t^2}\right) - \vec{\nabla}\left(\vec{\nabla} \cdot \vec{A} + \mu_0 \epsilon_0 \frac{\partial \phi}{\partial t}\right)} \tag{2.11}$$

The potential formulation has thus been achieved with those 2 equations. The 2nd equation is much more complicated than the magnetostatic version. We need to simplify the equations. This can be achieved by using the gauge degree of freedom that these potentials possess. A choice of gauge will be made which, hopefully simplifies these equations so that they can be solved.

2.2 Gauge transformations

Note that $\vec{E}$ and $\vec{B}$ are the physical fields and the ϕ and $\vec{A}$ potential fields are not. This gauge freedom exists because, according to the equations of electrodynamics, we can have different ϕ and $\vec{A}$ and yet the same $\vec{E}$ and $\vec{B}$ fields are obtained.[2] Let's work this out precisely.

$$\text{suppose, } \vec{A}' = \vec{A} + \vec{\alpha} \text{ and } \phi' = \phi + \beta \tag{2.12}$$
$$\text{take curl, } \vec{\nabla} \times \vec{A}' = \vec{\nabla} \times \vec{A} + \vec{\nabla} \times \vec{\alpha} \tag{2.13}$$
$$| \text{ the same } \vec{B} \text{ field is required}$$
$$\vec{B} = \vec{B} + \vec{\nabla} \times \vec{\alpha} \tag{2.14}$$
$$\vec{\nabla} \times \vec{\alpha} = 0 \tag{2.15}$$
$$\implies \vec{\alpha} = \vec{\nabla}\lambda \tag{2.16}$$

where λ is a general function called the "gauge-fixing function".

Now use,

$$\vec{E} = -\vec{\nabla}\phi - \frac{\partial \vec{A}}{\partial t} \tag{2.17}$$
$$\vec{E} = -\vec{\nabla}(\phi' - \beta) - \frac{\partial}{\partial t}(\vec{A}' - \vec{\alpha}) \tag{2.18}$$
$$\vec{E} = -\vec{\nabla}\phi' - \frac{\partial \vec{A}'}{\partial t} + \vec{\nabla}\beta + \frac{\partial \vec{\alpha}}{\partial t} \tag{2.19}$$
$$| \text{ we require } -\vec{\nabla}\phi' - \frac{\partial \vec{A}'}{\partial t} = \vec{E}, \text{ so}$$
$$0 = \vec{\nabla}\beta + \frac{\partial \vec{\alpha}}{\partial t} \tag{2.20}$$

---

[2]This is very unfamiliar to you because things in physics are usually defined to be physical but ϕ and $\vec{A}$ are not. They are mathematical objects.

Potential Formulation of Electrodynamics

$$0 = \vec{\nabla}\beta + \frac{\partial \vec{\nabla}\lambda}{\partial t} \tag{2.21}$$

$$0 = \vec{\nabla}\left(\beta + \frac{\partial \lambda}{\partial t}\right) \tag{2.22}$$

$$\beta = -\frac{\partial \lambda}{\partial t} \tag{2.23}$$

Thus,

$$\boxed{\vec{A}' = \vec{A} + \vec{\nabla}\lambda \text{ and } \phi' = \phi - \frac{\partial \lambda}{\partial t}} \tag{2.24}$$

represent the allowed gauge transformations[3] of ϕ and $\vec{A}$.

2.2.1 Coulomb gauge

We will now choose the so-called Coulomb gauge and see what happens to the potential formulation of electrodynamics.

Take $\vec{\nabla} \cdot \vec{A}' = \vec{\nabla} \cdot \vec{A} + \vec{\nabla} \cdot (\vec{\nabla}\lambda)$ and choose λ such that $\vec{\nabla} \cdot \vec{A}' = 0$ (Coulomb gauge).

$$\vec{\nabla}^2 \phi + \frac{\partial}{\partial t}(\vec{\nabla} \cdot \vec{A}) = -\frac{\rho}{\epsilon_0} \tag{2.25}$$

$$\vec{\nabla}^2 \left(\phi' + \frac{\partial \lambda}{\partial t}\right) + \frac{\partial}{\partial t}\left(\vec{\nabla} \cdot \vec{A}' - \vec{\nabla} \cdot (\vec{\nabla}\lambda)\right) = -\frac{\rho}{\epsilon_0} \tag{2.26}$$

| so the λ related terms cancel

$$\vec{\nabla}^2 \phi' + \frac{\partial}{\partial t}(\vec{\nabla} \cdot \vec{A}') = -\frac{\rho}{\epsilon_0} \tag{2.27}$$

| recall $\underbrace{\vec{\nabla} \cdot \vec{A}' = 0}_{\text{Coulomb gauge}}$

| drop the primes

$$\vec{\nabla}^2 \phi = -\frac{\rho}{\epsilon_0} \tag{2.28}$$

which is just Poisson equation. The solution is already known in electrostatics (section 16.7.4). However in electrodynamics, we also need $\vec{A}$ to get $\vec{E}$, so now we look at the other potential equation.

$$\left(\vec{\nabla}^2 \vec{A} - \mu_0 \epsilon_0 \frac{\partial^2 \vec{A}}{\partial t^2}\right) - \vec{\nabla}\left(\vec{\nabla} \cdot \vec{A} + \mu_0 \epsilon_0 \frac{\partial \phi}{\partial t}\right) = -\mu_0 \vec{J}$$

$$\left(\vec{\nabla}^2 (\vec{A}' - \vec{\nabla}\lambda) - \mu_0 \epsilon_0 \frac{\partial^2}{\partial t^2}(\vec{A}' - \vec{\nabla}\lambda)\right)$$
$$- \vec{\nabla}\left(\vec{\nabla} \cdot (\vec{A}' - \vec{\nabla}\lambda) + \mu_0 \epsilon_0 \frac{\partial}{\partial t}\left(\phi' + \frac{\partial \lambda}{\partial t}\right)\right) = -\mu_0 \vec{J}$$

note $\vec{\nabla}^2(\vec{\nabla}\lambda) = \vec{\nabla}(\vec{\nabla} \cdot \vec{\nabla}\lambda) - \vec{\nabla} \times (\vec{\nabla} \times (\vec{\nabla}\lambda))$ and $\vec{\nabla} \times (\vec{\nabla}\lambda)) = 0$ |

recall that $\vec{\nabla} \cdot \vec{A}'$ is chosen to be zero |

$$\vec{\nabla}^2 \vec{A}' - \mu_0 \epsilon_0 \frac{\partial^2 \vec{A}'}{\partial t^2} - \mu_0 \epsilon_0 \vec{\nabla}\left(\frac{\partial \phi'}{\partial t}\right) = -\mu_0 \vec{J} \tag{2.29}$$

---

[3]In Physics, we also call this an internal symmetry transformation.

where we can drop the prime labels also. Thus solving for $\vec{A}$ is very complicated, so Coulomb gauge does not look useful in getting solutions in electrodynamics.

2.2.2 Lorenz gauge

We again take $\vec{\nabla} \cdot \vec{A}' = \vec{\nabla} \cdot \vec{A} + \vec{\nabla} \cdot (\vec{\nabla} \lambda)$ and choose λ such that $\vec{\nabla} \cdot \vec{A}' = -\mu_0 \epsilon_0 \frac{\partial \phi'}{\partial t}$. This is the Lorenz gauge. Thus the potential equations become,[4]

$$\vec{\nabla}^2 \phi' + \frac{\partial}{\partial t}(\vec{\nabla} \cdot \vec{A}') = -\frac{\rho}{\epsilon_0} \tag{2.30}$$

$$\boxed{\vec{\nabla}^2 \phi' - \mu_0 \epsilon_0 \frac{\partial^2 \phi'}{\partial t^2} = -\frac{\rho}{\epsilon_0}} \tag{2.31}$$

and,

$$\boxed{\vec{\nabla}^2 \vec{A}' - \mu_0 \epsilon_0 \frac{\partial^2 \vec{A}'}{\partial t^2} = -\mu_0 \vec{J}} \tag{2.32}$$

This gauge has the virtue that it decouples ϕ and $\vec{A}$ and treats them on an equal footing. In Part II, we will also see that this gauge is Lorentz (not Lorenz!) invariant (equation (8.101)). In the next section, we will solve the potential equations in this gauge.

2.3 Solutions are retarded potentials

Recall the potential equations in Lorenz gauge (dropping the primes),

$$\vec{\nabla}^2 \phi - \mu_0 \epsilon_0 \frac{\partial^2 \phi}{\partial t^2} = -\frac{\rho}{\epsilon_0} \text{ and } \vec{\nabla}^2 \vec{A} - \mu_0 \epsilon_0 \frac{\partial^2 \vec{A}}{\partial t^2} = -\mu_0 \vec{J} \tag{2.33}$$

This is the problem of solving an inhomogeneous partial differential equation and we will do it using the method of Green's function.

The Green's function is defined as the inverse of the differential operator,

$$\underbrace{\left(\vec{\nabla}^2 - \mu_0 \epsilon_0 \frac{\partial^2}{\partial t^2}\right)}_{\text{think of this as the "coefficient matrix" in solving algebraic equation}} \underbrace{G(\vec{r}, t; \vec{r}', t')}_{\text{this is like the "inverse coefficient matrix"}} = \overbrace{\delta^3(\vec{r} - \vec{r}')\delta(t - t')}^{\text{continuous version of identity matrix}} \tag{2.34}$$

so the solution[5] is

$$\phi(\vec{r}, t) = -\frac{1}{\epsilon_0} \int G(\vec{r}, t; \vec{r}', t') \rho(\vec{r}', t') dV' dt' \tag{2.35}$$

---

[4]Historically, this gauge was wrongly attributed to H. A. Lorentz.
[5]Think of this as linear algebra where inverting the coefficient matrix will solve the unknown variables.

Potential Formulation of Electrodynamics

It is easy to see that when we apply $\left(\vec{\nabla}^2 - \mu_0\epsilon_0 \frac{\partial^2}{\partial t^2}\right)$, we get back the differential equation

$$\left(\vec{\nabla}^2 - \mu_0\epsilon_0 \frac{\partial^2}{\partial t^2}\right)\phi(\vec{r},t) = -\frac{1}{\epsilon_0}\int \left(\vec{\nabla}^2 - \mu_0\epsilon_0 \frac{\partial^2}{\partial t^2}\right)G(\vec{r},t;\vec{r}',t')\rho(\vec{r}',t')dV'dt' \quad (2.36)$$

$$\left(\vec{\nabla}^2 - \mu_0\epsilon_0 \frac{\partial^2}{\partial t^2}\right)\phi(\vec{r},t) = -\frac{1}{\epsilon_0}\int \delta^3(\vec{r}-\vec{r}')\delta(t-t')\rho(\vec{r}',t')dV'dt' \quad (2.37)$$

$$\left(\vec{\nabla}^2 - \mu_0\epsilon_0 \frac{\partial^2}{\partial t^2}\right)\phi(\vec{r},t) = -\frac{\rho(\vec{r},t)}{\epsilon_0} \quad (2.38)$$

We shall solve $G(\vec{r},t;\vec{r}',t')$ in Fourier space.[6]

$$\left(\vec{\nabla}^2 - \mu_0\epsilon_0 \frac{\partial^2}{\partial t^2}\right)G(\vec{r},t;\vec{r}',t') = \delta^3(\vec{r}-\vec{r}')\delta(t-t')$$

use Fourier form: $\delta^3(\vec{r}-\vec{r}') = \int_{-\infty}^{\infty} \frac{d^3\vec{k}}{(2\pi)^3}e^{i\vec{k}\cdot(\vec{r}-\vec{r}')}$ |

use Fourier form: $\delta(t-t') = \int_{-\infty}^{\infty} \frac{d\omega}{2\pi}e^{-i\omega(t-t')}$ |

use Fourier form: $G = \int_{-\infty}^{\infty} d^3\vec{k} \int_{-\infty}^{\infty} d\omega g(\vec{k},\omega)e^{i\vec{k}\cdot(\vec{r}-\vec{r}')}e^{-i\omega(t-t')}$ |

note this derivative $\begin{aligned}&\left(\vec{\nabla}^2 - \mu_0\epsilon_0 \frac{\partial^2}{\partial t^2}\right)e^{i\vec{k}\cdot(\vec{r}-\vec{r}')}e^{-i\omega(t-t')} \\ &= \left(-\vec{k}^2 + \frac{\omega^2}{c^2}\right)e^{i\vec{k}\cdot(\vec{r}-\vec{r}')}e^{-i\omega(t-t')}\end{aligned}$ |

where $\mu_0\epsilon_0 = \frac{1}{c^2}$, then comparing Fourier coefficients will give, |

$$g(\vec{k},\omega) = -\frac{1}{(2\pi)^4}\frac{1}{k^2 - \frac{\omega^2}{c^2}}$$

[7,8]where we wrote $\vec{k}^2 = k^2$ for simplicity. Hence $G(\vec{r},t;\vec{r}',t')$ is "solved".

$$G(\vec{r},t;\vec{r}',t') = -\frac{1}{(2\pi)^4}\int_{-\infty}^{\infty} d^3\vec{k} \int_{-\infty}^{\infty} d\omega \frac{1}{k^2 - \frac{\omega^2}{c^2}}e^{i\vec{k}\cdot(\vec{r}-\vec{r}')}e^{-i\omega(t-t')} \quad (2.39)$$

Well, we still have to carry out the integrals. The integrand has 2 simple poles at $\omega = \pm ck$. In order to carry out the integral with poles on the real axis, we shall treat ω as complex,[9] so $\omega = \omega_R + i\omega_I$ and use Cauchy's residue theorem: $\oint f(z)dz = 2\pi i \sum \text{Residues}$.

Therefore we need to choose a closed contour for the integration and making a choice of a closed contour amounts to specifying a (physical) boundary condition for the Green's function, as we shall see.

---

[6]Differential equations become algebraic under Fourier transform, so hopefully it becomes simpler.

[7]In Quantum Electrodynamics, this $g(\vec{k},\omega)$ the photon propagator in Lorenz gauge in momentum space.

[8]The choice of different signs in the temporal and spatial Fourier forms is to cater for the Minkowski metric in Special Relativity.

[9]The 2 poles prevented the integral to be done on the real line so we need a bigger space to do the integral.

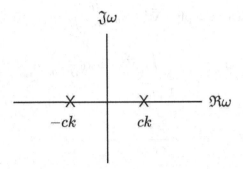

FIGURE 2.1
Two simple poles of the integrand.

Choice 1 (Retarded Boundary Condition):
The choice to skip above the poles is equivalent to moving the poles infinitesimally into the lower half plane (LHP). This is equivalent to rewriting G as

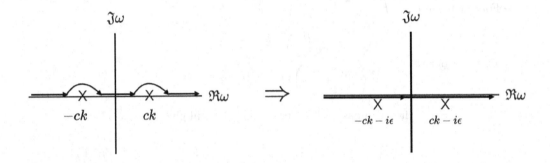

FIGURE 2.2
Skipping above the poles is equivalent to moving the poles down.

$$G(\vec{r},t;\vec{r}',t') = G^R(\vec{r},t;\vec{r}',t') \qquad (2.40)$$

$$G^R(\vec{r},t;\vec{r}',t') = \lim_{\epsilon \to 0} \frac{-1}{(2\pi)^4} \int_{-\infty}^{\infty} d^3\vec{k} \int_{-\infty}^{\infty} d\omega \frac{1}{k^2 - \frac{(\omega+i\epsilon)^2}{c^2}} e^{i\vec{k}\cdot(\vec{r}-\vec{r}')} e^{-i\omega(t-t')} \qquad (2.41)$$

| note $(ck)^2 - (\omega+i\epsilon)^2 = (ck - \omega - i\epsilon)(ck + \omega + i\epsilon)$

$$G^R(\vec{r},t;\vec{r}',t') = \lim_{\epsilon \to 0} \frac{1}{(2\pi)^4} \int_{-\infty}^{\infty} d^3\vec{k} \int_{-\infty}^{\infty} d\omega \frac{e^{i\vec{k}\cdot(\vec{r}-\vec{r}')} e^{-i\omega(t-t')} c^2}{(\omega + ck + i\epsilon)(\omega - ck + i\epsilon)} \qquad (2.42)$$

To carry out the complex integral, we need to close this path. If we close it in

- Upper Half Plane (UHP):
  - the closed path does not enclose any poles so the integral is zero
  - in UHP, $\omega_I > 0$ but we want $e^{-i\omega(t-t')} = e^{-i\omega_R(t-t')} e^{\omega_I(t-t')}$ to be damping[10] so we conclude that we need $t - t' < 0$ for $e^{\omega_I(t-t')}$ to be damping.
  - This means $G^R(\vec{r},t;\vec{r}',t') = 0$ for $t - t' < 0$.

---

[10] Damped integral means negative exponential $e^{-\alpha t}$ type.

Potential Formulation of Electrodynamics

- Lower Half Plane (LHP):
  - the closed path encloses 2 poles so the integral $\neq 0$
  - in LHP, $\omega_I < 0$ but we want $e^{-i\omega(t-t')} = e^{-i\omega_R(t-t')}e^{\omega_I(t-t')}$ to be damping so we conclude that we need $t - t' > 0$ for $e^{\omega_I(t-t')}$ to be damping.

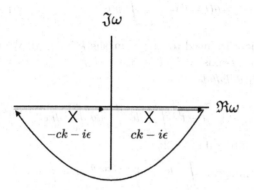

FIGURE 2.3
Closing the contour over the lower half plane (LHP).

Now we calculate explicitly the result when we close this path over LHP. Closing over LHP, where $\omega_I < 0$, the damping of the integral (due to $e^{\omega_I(t-t')}$) requires $t - t' > 0$. So now we can evaluate the contour integral,

$$\oint d\omega \frac{e^{-i\omega(t-t')}c^2}{(\omega + ck + i\epsilon)(\omega - ck + i\epsilon)} = -2\pi i \sum \text{Residues} \tag{2.43}$$

| negative sign is due to the clockwise closed contour
| break $\oint d\omega$ into $\int_{\text{straight}} d\omega + \int_{\text{curve}} d\omega$
| then $\int_{\text{curve}} d\omega \to 0$ due to damping and $\omega_I \to -\infty$
| and $\int_{\text{straight}} d\omega = \int_{-\infty}^{\infty} d\omega$

$$\int_{-\infty}^{\infty} d\omega \frac{e^{-i\omega(t-t')}c^2}{(\omega + ck + i\epsilon)(\omega - ck + i\epsilon)} = -2\pi i c^2 \left[\frac{e^{-i(-ck-i\epsilon)(t-t')}}{-ck - i\epsilon - ck + i\epsilon} + \frac{e^{-i(ck-i\epsilon)(t-t')}}{ck - i\epsilon + ck + i\epsilon} \right]$$

| take the limit $\epsilon \to 0$

$$\int_{-\infty}^{\infty} d\omega \frac{e^{-i\omega(t-t')}c^2}{(\omega + ck + i\epsilon)(\omega - ck + i\epsilon)} = \frac{-2\pi i c^2}{2ck} \left(-e^{ick(t-t')} + e^{-ick(t-t')} \right) \tag{2.44}$$

$$\int_{-\infty}^{\infty} d\omega \frac{e^{-i\omega(t-t')}c^2}{(\omega + ck + i\epsilon)(\omega - ck + i\epsilon)} = \frac{-\pi i c}{k} (-2i\sin(ck(t-t'))) \tag{2.45}$$

$$\int_{-\infty}^{\infty} d\omega \frac{e^{-i\omega(t-t')}c^2}{(\omega + ck + i\epsilon)(\omega - ck + i\epsilon)} = -\frac{2\pi c}{k} \sin(ck(t-t')) \tag{2.46}$$

In summary, we have

$$G^R(\vec{r},t;\vec{r}',t') = \begin{cases} 0 & t-t' < 0 \\ -\frac{c}{(2\pi)^3} \int d^3\vec{k} \frac{e^{i\vec{k}\cdot(\vec{r}-\vec{r}')}}{k} \sin(ck(t-t')) & t-t' > 0 \end{cases} \quad (2.47)$$

| we can combine both cases using the step function $\theta(t-t')$

$$G^R(\vec{r},t;\vec{r}',t') = -\theta(t-t')\frac{c}{(2\pi)^3} \int d^3\vec{k} \frac{e^{i\vec{k}\cdot(\vec{r}-\vec{r}')}}{k} \sin(ck(t-t')) \quad (2.48)$$

To finish the calculation, we need to carry out the k integral. We use spherical coordinates and align $\vec{r}-\vec{r}'$ to the z-axis so $\vec{k}\cdot(\vec{r}-\vec{r}') = k|\vec{r}-\vec{r}'|\cos\theta = kR\cos\theta$ and the volume element is $d^3\vec{k} = k^2 \sin\theta dk d\theta d\phi$.

$$G^R(\vec{r},t;\vec{r}',t') = -\frac{c}{(2\pi)^3}\theta(t-t') \int_0^{2\pi} d\phi \int_0^{\pi} d\theta \int_0^{\infty} dk\, e^{ikR\cos\theta} k\sin\theta \sin(ck(t-t'))$$

| ϕ integral gives 2π

| θ integral: $\int_0^{\pi} d\theta \sin\theta e^{ikR\cos\theta} = -\int_0^{\pi} d\cos\theta\, e^{ikR\cos\theta} = \frac{2}{kR}\sin(kR)$

$$G^R(\vec{r},t;\vec{r}',t') = -\frac{2c}{(2\pi)^2 R}\theta(t-t') \int_0^{\infty} dk \sin(kR)\sin(ck(t-t')) \quad (2.49)$$

| write $\sin(kR) = \frac{1}{2i}(e^{ikR} - e^{-ikR})$ and similarly for $\sin(ck(t-t'))$

$$G^R(\vec{r},t;\vec{r}',t') = \theta(t-t')\frac{2c}{8\pi R}\int_0^{\infty} \frac{dk}{2\pi}\Big(e^{ik(R+c(t-t'))} + e^{-ik(R+c(t-t'))}$$
$$-e^{ik(R-c(t-t'))} - e^{-ik(R-c(t-t'))}\Big) \quad (2.50)$$

| note 2nd term: $\int_0^{\infty} \frac{dk}{2\pi} e^{-ik(R+c(t-t'))} \xrightarrow{k_1=-k} -\int_0^{-\infty} \frac{dk_1}{2\pi} e^{ik_1(R+c(t-t'))}$

| then $-\int_0^{-\infty} = \int_{-\infty}^0$ so the 2nd term adds to 1st term.

| The 4th term adds similarly to the 3rd term

$$G^R(\vec{r},t;\vec{r}',t') = \theta(t-t')\frac{c}{4\pi R}\int_{-\infty}^{\infty} \frac{dk}{2\pi}\left[e^{ik(R+c(t-t'))} - e^{ik(R-c(t-t'))}\right] \quad (2.51)$$

| note: $\underbrace{\int_{-\infty}^{\infty} \frac{dk}{2\pi} e^{ik(R+c(t-t'))} = \delta(R+c(t-t'))}_{\text{Fourier form of the delta function}}$

| then use identity: $c\delta(R+c(t-t')) = \delta\left(t-t'+\frac{R}{c}\right)$

$$G^R(\vec{r},t;\vec{r}',t') = \theta(t-t')\frac{1}{4\pi R}\left[\delta\left(t-t'+\frac{R}{c}\right) - \delta\left(-t+t'+\frac{R}{c}\right)\right] \quad (2.52)$$

| recall that the delta function is even
| drop first delta function as we require $t-t' > 0$

$$G^R(\vec{r},t;\vec{r}',t') = -\theta(t-t')\frac{1}{4\pi R}\delta\left(t-t'-\frac{R}{c}\right) \quad (2.53)$$

Potential Formulation of Electrodynamics

Choice 2 (Advanced Boundary Condition):

This will just be briefly mentioned as the steps are very similar and this is an unphysical solution. Skipping over the poles from below means,

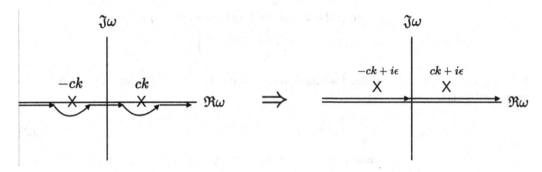

FIGURE 2.4
Skipping below the poles is equivalent to moving the poles up.

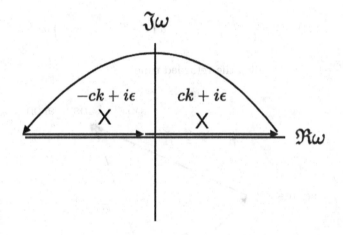

FIGURE 2.5
Closing the contour over the upper half plane (UHP).

$$G(\vec{r},t;\vec{r}',t') = G^A(\vec{r},t;\vec{r}',t') \qquad (2.54)$$

$$= \lim_{\epsilon \to 0} -\frac{1}{(2\pi)^4} \int d^3\vec{k} \int d\omega \frac{1}{k^2 - \frac{(\omega - i\epsilon)^2}{c^2}} e^{i\vec{k}\cdot(\vec{r}-\vec{r}')} e^{-i\omega(t-t')} \qquad (2.55)$$

For the same reasons of damping the integral, $G^A(\vec{r},t;\vec{r}',t') = 0$ for $t - t' > 0$. The path is to be closed over UHP ($\omega_I > 0$).

After evaluating the contour integral, we have in summary, [11]

$$G^A(\vec{r},t;\vec{r}',t') = \begin{cases} 0 & t - t' > 0 \\ \frac{c}{(2\pi)^3} \int d^3\vec{k} \frac{e^{i\vec{k}\cdot(\vec{r}-\vec{r}')}}{k} \sin(ck(t-t')) & t - t' < 0 \end{cases} \qquad (2.56)$$

$$\text{or, } G^A(\vec{r},t;\vec{r}',t') = \theta(t'-t)\frac{c}{(2\pi)^3} \int d^3\vec{k} \frac{e^{i\vec{k}\cdot(\vec{r}-\vec{r}')}}{k} \sin(ck(t-t')) \qquad (2.57)$$

---

[11] The "missing" negative sign is due to the contour being closed in an anticlockwise manner.

Carry out the k integral in the same way to get,

$$G^A(\vec{r},t;\vec{r}',t') = -\theta(t'-t)\frac{1}{4\pi R}\left[\delta\left(t-t'+\frac{R}{c}\right)-\delta\left(t-t'-\frac{R}{c}\right)\right] \quad (2.58)$$

| drop the 2nd delta function as we require $t-t'<0$

$$G^A(\vec{r},t;\vec{r}',t') = -\theta(t'-t)\frac{1}{4\pi R}\delta\left(t-t'+\frac{R}{c}\right) \quad (2.59)$$

Ignoring the unphysical advanced solution and keeping the retarded solution leads us to the scalar potential

$$\phi(\vec{r},t) = -\frac{1}{\epsilon_0}\int G^R(\vec{r},t;\vec{r}',t')\rho(\vec{r}',t')dV'dt' \quad (2.60)$$

| usually $\theta(t-t')$ is left out and recall $R=|\vec{r}-\vec{r}'|$

$$\phi(\vec{r},t) = \frac{1}{4\pi\epsilon_0}\int\int\frac{\rho(\vec{r}',t')}{|\vec{r}-\vec{r}'|}\delta\left(t-t'-\frac{|\vec{r}-\vec{r}'|}{c}\right)dV'dt' \quad (2.61)$$

Similarly, for the vector potential, we have

$$\vec{A}(\vec{r},t) = \frac{\mu_0}{4\pi}\int\int\frac{\vec{J}(\vec{r}',t')}{|\vec{r}-\vec{r}'|}\delta\left(t-t'-\frac{|\vec{r}-\vec{r}'|}{c}\right)dV'dt' \quad (2.62)$$

where $t' = t - \frac{|\vec{r}-\vec{r}'|}{c} = t_r$ is called the retarded time.

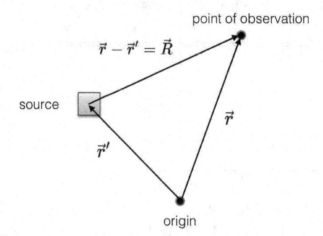

FIGURE 2.6
The information at the point of observation at time t is from the source at the retarded time t_r.

To understand retarded time, we see that $\frac{|\vec{r}-\vec{r}'|}{c}$ is the time for "information" to travel at speed c from the source to the point of observation. Thus the potential at time t is due to the "information" from the source at t_r. This is a clue that electrodynamics and relativity are related as relativity requires causality where information takes a finite speed (maximum is c) to travel.

2.4 The fields: Jefimenko's equations

Note that it appears that $\phi(\vec{r}, t)$ and $\vec{A}(\vec{r}, t)$ differs from their electrostatic and magnetostatic versions just by the retarded time. It is not that simple. If we take the electrostatic and magnetostatic versions of the fields and simply put in the retarded time, the results are wrong.

$$\vec{E}(\vec{r}) = \frac{1}{4\pi\epsilon_0} \int \frac{\rho(\vec{r}')}{|\vec{r} - \vec{r}'|^2} \hat{R} dV' \tag{2.63}$$

$$\implies \vec{E}(\vec{r}, t) \neq \frac{1}{4\pi\epsilon_0} \int \int \frac{\rho(\vec{r}', t')}{|\vec{r} - \vec{r}'|^2} \hat{R} \delta\left(t - t' - \frac{|\vec{r} - \vec{r}'|}{c}\right) dV' dt' \tag{2.64}$$

$$\vec{B}(\vec{r}) = \frac{\mu_0}{4\pi} \int \frac{\vec{J}(r') \times \hat{R}}{|\vec{r} - \vec{r}'|^2} dV' \tag{2.65}$$

$$\implies \vec{B}(\vec{r}, t) \neq \frac{\mu_0}{4\pi} \int \int \frac{\vec{J}(\vec{r}', t') \times \hat{R}}{|\vec{r} - \vec{r}'|^2} \delta\left(t - t' - \frac{|\vec{r} - \vec{r}'|}{c}\right) dV' dt' \tag{2.66}$$

To get the fields, we need to use the retarded potentials and the equations: $\vec{E} = -\vec{\nabla}\phi - \frac{\partial \vec{A}}{\partial t}$ and $\vec{B} = \vec{\nabla} \times \vec{A}$. We do the easier $\frac{\partial \vec{A}}{\partial t}$ term first,

$$\frac{\partial \vec{A}}{\partial t} = \frac{\partial}{\partial t} \frac{\mu_0}{4\pi} \int \frac{\vec{J}(\vec{r}', t_r)}{|\vec{r} - \vec{r}'|} dV' = \frac{\mu_0}{4\pi} \int \frac{\dot{\vec{J}}(\vec{r}', t_r)}{|\vec{r} - \vec{r}'|} dV' \tag{2.67}$$

Then we calculate $\vec{\nabla}\phi$ but note that $\vec{r}$ is in the denominator as $|\vec{r} - \vec{r}'|$ and is hiding in $t_r = t - \frac{|\vec{r} - \vec{r}'|}{c}$.

$$\vec{\nabla}\phi(\vec{r}, t) = \frac{1}{4\pi\epsilon_0} \int \left(\frac{\vec{\nabla}\rho(\vec{r}', t_r)}{|\vec{r} - \vec{r}'|} + \rho(\vec{r}', t_r) \vec{\nabla} \frac{1}{|\vec{r} - \vec{r}'|} \right) dV' \tag{2.68}$$

$$\mid \text{ write } \vec{\nabla}\rho = \frac{\partial \rho}{\partial t} \frac{\partial t}{\partial t_r} \vec{\nabla} t_r = \frac{\partial \rho}{\partial t}\left(-\frac{\vec{\nabla}|\vec{r} - \vec{r}'|}{c}\right) = -\frac{1}{c}\dot{\rho}\frac{\vec{r} - \vec{r}'}{|\vec{r} - \vec{r}'|}$$

$$\mid \text{ and write } \vec{\nabla}\frac{1}{|\vec{r} - \vec{r}'|} = -\frac{\vec{r} - \vec{r}'}{|\vec{r} - \vec{r}'|^3} = -\frac{\hat{R}}{|\vec{r} - \vec{r}'|^2}$$

$$\vec{\nabla}\phi(\vec{r}, t) = \frac{1}{4\pi\epsilon_0} \int \left(-\frac{\dot{\rho}}{c} \frac{\hat{R}}{|\vec{r} - \vec{r}'|} - \rho \frac{\hat{R}}{|\vec{r} - \vec{r}'|^2} \right) dV' \tag{2.69}$$

$$\vec{E}(\vec{r}, t) = -\vec{\nabla}\phi - \frac{\partial \vec{A}}{\partial t} \tag{2.70}$$

$$\vec{E}(\vec{r}, t) = \frac{1}{4\pi\epsilon_0} \int \left(\frac{\rho(\vec{r}', t_r)}{|\vec{r} - \vec{r}'|^2} \hat{R} + \frac{\dot{\rho}(\vec{r}', t_r)}{c|\vec{r} - \vec{r}'|} \hat{R} - \frac{\dot{\vec{J}}(\vec{r}', t_r)}{c^2 |\vec{r} - \vec{r}'|} \right) dV' \tag{2.71}$$

So there are 2 more terms extra as compared to our naive guess. This Jefimenko's equation for the E-field is essentially the time-dependent generalisation of Coulomb's law.

Now for the B-field,

$$\vec{B}(\vec{r},t) = \vec{\nabla} \times \vec{A} \qquad (2.72)$$

| recall $\vec{\nabla} \times (f\vec{A}) = f(\vec{\nabla} \times \vec{A}) - \vec{A} \times (\vec{\nabla} f)$

$$\vec{B}(\vec{r},t) = \frac{\mu_0}{4\pi} \int \left(\frac{\vec{\nabla} \times \vec{J}}{|\vec{r} - \vec{r}'|} - \vec{J} \times \vec{\nabla} \frac{1}{|\vec{r} - \vec{r}'|} \right) dV' \qquad (2.73)$$

| we already know $\vec{\nabla} \dfrac{1}{|\vec{r} - \vec{r}'|} = -\dfrac{\vec{r} - \vec{r}'}{|\vec{r} - \vec{r}'|^3} = -\dfrac{\hat{R}}{|\vec{r} - \vec{r}'|^2}$

| look at $(\vec{\nabla} \times \vec{J})_x = \dfrac{\partial J_z}{\partial y} - \dfrac{\partial J_y}{\partial z} = \dfrac{\partial J_z}{\partial t} \dfrac{\partial t}{\partial t_r} \dfrac{\partial t_r}{\partial y} - \dfrac{\partial J_y}{\partial t} \dfrac{\partial t}{\partial t_r} \dfrac{\partial t_r}{\partial z} = \dot{J}_z \dfrac{\partial t_r}{\partial y} - \dot{J}_y \dfrac{\partial t_r}{\partial z}$

| recall that $\vec{\nabla} t_r = \dfrac{\vec{r} - \vec{r}'}{|\vec{r} - \vec{r}'|} \left(-\dfrac{1}{c}\right) = -\dfrac{1}{c}\hat{R}$

| hence, $(\vec{\nabla} \times \vec{J})_x = \dot{J}_z \left(-\dfrac{1}{c}\hat{R}_y\right) - \dot{J}_y \left(-\dfrac{1}{c}\hat{R}_z\right) = \dfrac{1}{c}(\dot{\vec{J}} \times \hat{R})_x$

$$\vec{B}(\vec{r},t) = \frac{\mu_0}{4\pi} \int \left(\frac{\dot{\vec{J}}(\vec{r}',t_r) \times \hat{R}}{c|\vec{r} - \vec{r}'|} + \frac{\vec{J}(\vec{r}',t_r) \times \hat{R}}{|\vec{r} - \vec{r}'|^2} \right) dV' \qquad (2.74)$$

This Jefimenko's equation for B-field is essentially the time-dependent generalisation of Biot-Savart law.[12]

These 2 Jefimenko's equations are essentially formal and have limited uses because of the difficult integral due to t_r having dependence on $\vec{r}'$. However, they complete the description of electrodynamics in vector calculus language before we recast into a relativistic description.

---

[12]Caution: This means that formulas such as mutual inductance derived in year 2 EM should be taken with a pinch of salt.

3

Standard Application 1: Point Charge

3.1 Retarded potentials: Lienard–Wiechert potentials

Now apply the retarded potentials $\phi(\vec{r},t)$ and $\vec{A}(\vec{r},t)$ to a moving charge of charge q with the following notation:

- Point charge trajectory: $\vec{r}_0(t)$
- Point charge velocity: $\vec{v}(t) = \frac{d\vec{r}_0(t)}{dt}$
- Point charge density: $\rho(\vec{r},t) = q\delta^3(\vec{r} - \vec{r}_0(t))$
- Point charge current density: $\vec{J}(\vec{r},t) = q\vec{v}(t)\delta^3(\vec{r} - \vec{r}_0(t))$

$$\phi(\vec{r},t) = \frac{1}{4\pi\epsilon_0} \int\int \frac{\rho(\vec{r}',t')}{|\vec{r}-\vec{r}'|} \delta\left(t - t' - \frac{|\vec{r}-\vec{r}'|}{c}\right) dV' dt' \tag{3.1}$$

$$\phi(\vec{r},t) = \frac{q}{4\pi\epsilon_0} \int\int \frac{\delta^3(\vec{r}' - \vec{r}_0(t'))}{|\vec{r}-\vec{r}'|} \delta\left(t - t' - \frac{|\vec{r}-\vec{r}'|}{c}\right) dV' dt' \tag{3.2}$$

| use the delta function to do the volume integral

$$\phi(\vec{r},t) = \frac{q}{4\pi\epsilon_0} \int \frac{\delta\left(t - t' - \frac{|\vec{r}-\vec{r}_0(t')|}{c}\right)}{|\vec{r} - \vec{r}_0(t')|} dt' \tag{3.3}$$

| because the dependence on t' is implicit, we need an identity

| identity: $\int ds\, \delta(f(s))g(s) = \int ds \frac{df}{ds}\frac{\delta(f(s))}{\frac{df}{ds}}g(s) = \int df\, \delta(f) \frac{g(s)}{\frac{df}{ds}} = \left.\frac{g(s)}{\frac{df}{ds}}\right|_{f(s)=0}$

| so, $f(t') = t' - t + \frac{|\vec{r} - \vec{r}_0(t')|}{c}$

| $= t' - t + \frac{1}{c}\sqrt{(x-x_0(t'))^2 + (y-y_0(t'))^2 + (z-z_0(t'))^2}$

| so, $\frac{df(t')}{dt'} = 1 - \frac{(x-x_0(t'))\dot{x}_0(t') + \cdots}{c\sqrt{(x-x_0(t'))^2 + \cdots}}$

| $= 1 - \frac{(\vec{r} - \vec{r}_0(t'))\cdot\vec{v}(t')}{c|\vec{r}-\vec{r}_0(t')|} = 1 - \frac{\hat{R}(t')\cdot\vec{v}(t')}{c}$

$$\boxed{\phi(\vec{r},t) = \frac{qc}{4\pi\epsilon_0} \frac{1}{R(t')c - \vec{R}(t')\cdot\vec{v}(t')}\bigg|_{f(t')=0 \text{ or } t'=t-\frac{R(t')}{c}=t-\frac{|\vec{r}-\vec{r}_0(t')|}{c}}} \tag{3.4}$$

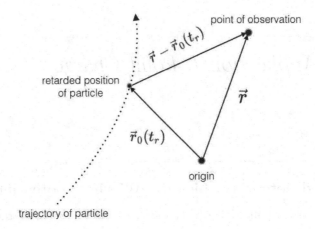

FIGURE 3.1
The setup to calculate the retarded potentials of a moving charge.

Then,

$$\vec{A}(\vec{r},t) = \frac{q\mu_0}{4\pi} \int\int \frac{\vec{v}(t')\delta^3(\vec{r}\,'-\vec{r}_0(t'))}{|\vec{r}-\vec{r}\,'|}\delta\left(t-t'-\frac{|\vec{r}-\vec{r}\,'|}{c}\right)dV'dt' \qquad (3.5)$$

| so the only difference here is that $g(s) = \dfrac{\vec{v}(t')}{|\vec{r}-\vec{r}_0(t')|}$, so

$$\vec{A}(\vec{r},t) = \frac{q\mu_0}{4\pi}\frac{cv(t')}{R(t')c - \vec{R}(t')\cdot\vec{v}(t')}\bigg|_{t'=t-\frac{R(t')}{c}=t-\frac{|\vec{r}-\vec{r}_0(t')|}{c}} \qquad (3.6)$$

$$\boxed{\vec{A}(\vec{r},t) = \frac{\vec{v}(t')}{c^2}\phi(\vec{r},t')\bigg|_{t'=t-\frac{R(t')}{c}=t-\frac{|\vec{r}-\vec{r}_0(t')|}{c}}} \qquad (3.7)$$

We can provide a nice physical (actually geometrical) picture to the derivation. We write,[1]

$$\phi(\vec{r},t) = \frac{1}{4\pi\epsilon_0}\int\int\frac{\rho(\vec{r}\,',t')}{|\vec{r}-\vec{r}\,'|}\delta(t_r-t')dV'dt' \qquad (3.8)$$

| do the time integral formally

$$\phi(\vec{r},t) = \frac{1}{4\pi\epsilon_0}\int\frac{\rho(\vec{r}\,',t_r)}{|\vec{r}-\vec{r}\,'|}dV' \qquad (3.9)$$

The picture is that, information comes to the point of observation at speed c and the charge is moving so $\int \rho dV'$ is not the total charge. This is a geometric effect and has nothing to do with relativity.

---

[1] If you find this rough derivation actually confusing then feel free to skip over it.

Standard Application 1: Point Charge

If the charge distribution is not moving:

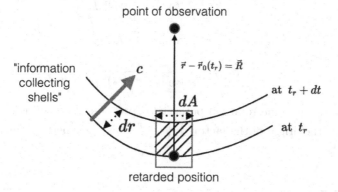

$$\text{charge "seen" if charge distribution is not moving} = \text{shaded area} = \rho dA dr \quad (3.10)$$
$$= \rho dV \quad (3.11)$$

If the charge distribution is moving:

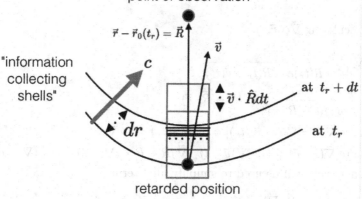

$$\text{Shifted area} = \rho((\vec{v} \cdot \hat{R})dt)dA \quad (3.12)$$
$$\text{where } dt = \frac{dr}{c}$$
$$\text{Shifted area} = \rho \vec{v} \cdot \hat{R} \frac{dr}{c} dA \quad (3.13)$$
$$\text{Shifted area} = \frac{\vec{v} \cdot \hat{R}}{c} \rho dV \quad \text{which is actually less}$$

charge "seen" if charge distribution is moving with velocity $\vec{v}$ = shaded area

$$dq = \rho dV - \frac{\vec{v} \cdot \hat{R}}{c} \rho dV$$

$$\int \rho dV = \int \frac{1}{1 - \frac{\vec{v} \cdot \hat{R}}{c}} dq$$

$$\text{so roughly,} \quad \phi \approx \frac{1}{4\pi\epsilon_0} \frac{1}{|\vec{r} - \vec{r}_0(t_r)|} \int \rho \, dV \tag{3.14}$$

$$\phi = \frac{1}{4\pi\epsilon_0} \frac{1}{|\vec{r} - \vec{r}_0(t_r)|} \int \frac{1}{1 - \frac{\vec{v} \cdot \hat{R}}{c}} dq \tag{3.15}$$

$$\phi = \frac{1}{4\pi\epsilon_0} \frac{1}{|\vec{r} - \vec{r}_0(t_r)|} \frac{1}{1 - \frac{\vec{v} \cdot \hat{R}}{c}} \int dq \tag{3.16}$$

and $\int dq = q$ gives the same retarded potential expression. Again, this rough argument is just to show that the origin of the factor: $1/(1 - \frac{\vec{v} \cdot \hat{R}}{c})$ is a geometrical one.

3.2 Fields

The equations for the fields are $\vec{E} = -\vec{\nabla}\phi - \frac{\partial \vec{A}}{\partial t}$ and $\vec{B} = \vec{\nabla} \times \vec{A}$. The main difficulty in calculation is that the differentiation needs to be done implicitly. Note the expressions

$$t_r = t - \frac{|\vec{r} - \vec{r}_0(t_r)|}{c} = t - \frac{R(t_r)}{c} \implies R(t_r) = c(t - t_r) \tag{3.17}$$

$$\text{and,} \quad \vec{v}(t) = \frac{d\vec{r}_0(t)}{dt} \tag{3.18}$$

We start by calculating $\vec{\nabla}\phi(\vec{r}, t)$,

$$\vec{\nabla}\phi(\vec{r}, t) = \vec{\nabla} \frac{qc}{4\pi\epsilon_0} \frac{1}{R(t_r)c - \vec{R}(t_r) \cdot \vec{v}(t_r)} \tag{3.19}$$

$$\vec{\nabla}\phi(\vec{r}, t) = \frac{qc}{4\pi\epsilon_0} \frac{-1}{(Rc - \vec{R} \cdot \vec{v})^2} \vec{\nabla}(Rc - \vec{R} \cdot \vec{v}) \tag{3.20}$$

| note $\vec{\nabla} R = \vec{\nabla}|\vec{r} - \vec{r}_0(t_r)| = \vec{\nabla} c(t - t_r) = -c\vec{\nabla} t_r$
| note $\vec{\nabla}(\vec{R} \cdot \vec{v}) = (\vec{R} \cdot \vec{\nabla})\vec{v} + (\vec{v} \cdot \vec{\nabla})\vec{R} + \vec{R} \times (\vec{\nabla} \times \vec{v}) + \vec{v} \times (\vec{\nabla} \times \vec{R})$
| chain rule will be used to simplify all 4 terms
| first term: $(\vec{R} \cdot \vec{\nabla})\vec{v} = \left(R_x \frac{\partial}{\partial x} + \cdots\right)\vec{v}(t_r) = R_x \frac{d\vec{v}(t_r)}{dt_r} \frac{\partial t_r}{\partial x} + \cdots = \vec{a}(\vec{R} \cdot \vec{\nabla} t_r)$
| second term: $(\vec{v} \cdot \vec{\nabla})\vec{R} = (\vec{v} \cdot \vec{\nabla})\vec{r} - (\vec{v} \cdot \vec{\nabla})\vec{r}_0(t_r) = \vec{v} - \vec{v}(\vec{v} \cdot \vec{\nabla} t_r)$
| third term: $\vec{R} \times (\vec{\nabla} \times \vec{v}) = \vec{R} \times (-\vec{a} \times \vec{\nabla} t_r)$
| fourth term: $\vec{v} \times (\vec{\nabla} \times \vec{R}) = \vec{v} \times (\vec{\nabla} \times \vec{r} - \vec{\nabla} \times \vec{r}_0(t_r)) = \vec{v} \times (\vec{v} \times \vec{\nabla} t_r)$

$$\vec{\nabla}\phi(\vec{r}, t) = \frac{qc}{4\pi\epsilon_0} \frac{-1}{(Rc - \vec{R} \cdot \vec{v})^2} \Big[-c^2 \vec{\nabla} t_r - \vec{a}(\vec{R} \cdot \vec{\nabla} t_r) - \vec{v} + \vec{v}(\vec{v} \cdot \vec{\nabla} t_r)$$
$$+ \vec{R} \times (\vec{a} \times \vec{\nabla} t_r) - \vec{v} \times (\vec{v} \times \vec{\nabla} t_r) \Big]$$

| use identity $\vec{R} \times (\vec{a} \times \vec{\nabla} t_r) - \vec{a}(\vec{R} \cdot \vec{\nabla} t_r) = -(\vec{R} \cdot \vec{a})\vec{\nabla} t_r$
| use identity $\vec{v}(\vec{v} \cdot \vec{\nabla} t_r) - \vec{v} \times (\vec{v} \times \vec{\nabla} t_r) = v^2 \vec{\nabla} t_r$

$$\vec{\nabla}\phi(\vec{r}, t) = \frac{qc}{4\pi\epsilon_0} \frac{1}{(Rc - \vec{R} \cdot \vec{v})^2} \left[\vec{v} + \left(c^2 - v^2 + (\vec{R} \cdot \vec{a})\right) \vec{\nabla} t_r \right] \tag{3.21}$$

| lastly, use $-c\vec{\nabla} t_r = \vec{\nabla} R = \vec{\nabla}\sqrt{\vec{R} \cdot \vec{R}} = \frac{\vec{\nabla}(\vec{R} \cdot \vec{R})}{2\sqrt{\vec{R} \cdot \vec{R}}} = \frac{(\vec{R} \cdot \vec{\nabla})\vec{R} + \vec{R} \times (\vec{\nabla} \times \vec{R})}{R}$

Standard Application 1: Point Charge

| then chain rule give $(\vec{R}\cdot\vec{\nabla})\vec{R} = \vec{R} - \vec{v}(\vec{R}\cdot\vec{\nabla}t_r)$ and $\vec{\nabla}\times\vec{R} = \vec{v}\times\vec{\nabla}t_r$

| then, $\vec{R}\times(\vec{v}\times\vec{\nabla}t_r) = \vec{v}(\vec{R}\cdot\vec{\nabla}t_r) - (\vec{R}\cdot\vec{v})\vec{\nabla}t_r$ and 2 terms cancel

| rearrange and make $\vec{\nabla}t_r$ the subject to get $\vec{\nabla}t_r = -\dfrac{\vec{R}}{Rc - \vec{R}\cdot\vec{v}}$

$$\vec{\nabla}\phi(\vec{r},t) = \frac{qc}{4\pi\epsilon_0}\frac{1}{(Rc - \vec{R}\cdot\vec{v})^2}\left[\vec{v} - \left(c^2 - v^2 + \vec{R}\cdot\vec{a}\right)\frac{\vec{R}}{Rc - \vec{R}\cdot\vec{v}}\right] \tag{3.22}$$

$$\vec{\nabla}\phi(\vec{r},t) = \frac{qc}{4\pi\epsilon_0}\frac{1}{(Rc - \vec{R}\cdot\vec{v})^3}\left[\left(Rc - \vec{R}\cdot\vec{v}\right)\vec{v} - \left(c^2 - v^2 + \vec{R}\cdot\vec{a}\right)\vec{R}\right] \tag{3.23}$$

By repeating the calculation in a similar way, $\frac{\partial \vec{A}}{\partial t}$ is obtained as

$$\frac{\partial \vec{A}}{\partial t} = \frac{qc}{4\pi\epsilon_0}\frac{1}{(Rc - \vec{R}\cdot\vec{v})^3}\left[\left(Rc - \vec{R}\cdot\vec{v}\right)\left(\frac{R\vec{a}}{c} - \vec{v}\right) + \left(c^2 - v^2 + \vec{R}\cdot\vec{a}\right)\frac{R\vec{v}}{c}\right] \tag{3.24}$$

So the electric field is

$$\vec{E}(\vec{r},t) = -\vec{\nabla}\phi(\vec{r},t) - \frac{\partial \vec{A}(\vec{r},t)}{\partial t} \tag{3.25}$$

$$\vec{E}(\vec{r},t) = \frac{qc}{4\pi\epsilon_0}\frac{1}{(Rc - \vec{R}\cdot\vec{v})^3}\left[\left(Rc - \vec{R}\cdot\vec{v}\right)\left(-\vec{v} - \frac{R\vec{a}}{c} + \vec{v}\right)\right.$$
$$\left. + \left(c^2 - v^2 + \vec{R}\cdot\vec{a}\right)\left(\vec{R} - \frac{R\vec{v}}{c}\right)\right]$$

$$\vec{E}(\vec{r},t) = \frac{q}{4\pi\epsilon_0}\frac{R}{(Rc - \vec{R}\cdot\vec{v})^3}\left[-(Rc - \vec{R}\cdot\vec{v})\vec{a} + \left(c^2 - v^2 + \vec{R}\cdot\vec{a}\right)(\hat{R}c - \vec{v})\right] \tag{3.26}$$

$$\vec{E}(\vec{r},t) = \frac{q}{4\pi\epsilon_0}\frac{R}{(Rc - \vec{R}\cdot\vec{v})^3}\left[(c^2 - v^2)(\hat{R}c - \vec{v}) + (\vec{R}\cdot\vec{a})\hat{R}c - (\vec{R}\cdot\vec{a})\vec{v} - Rc\vec{a} + (\vec{R}\cdot\vec{v})\vec{a}\right] \tag{3.27}$$

| we can check backwards that the last 4 terms become $\vec{R}\times((\hat{R}c - \vec{v})\times\vec{a})$

$$\boxed{\vec{E}(\vec{r},t) = \frac{q}{4\pi\epsilon_0}\frac{R}{(Rc - \vec{R}\cdot\vec{v})^3}\left[(c^2 - v^2)(\hat{R}c - \vec{v}) + \vec{R}\times((\hat{R}c - \vec{v})\times\vec{a})\right]} \tag{3.28}$$

- Term $\frac{q}{4\pi\epsilon_0}\frac{R}{(Rc - \vec{R}\cdot\vec{v})^3}(c^2 - v^2)(\hat{R}c - \vec{v})$ is called the velocity field term and it falls as $\approx \frac{1}{R^2}$.

- Term $\frac{q}{4\pi\epsilon_0}\frac{R}{(Rc - \vec{R}\cdot\vec{v})^3}\vec{R}\times((\hat{R}c - \vec{v})\times\vec{a})$ is called the acceleration field or the radiation field and it falls as $\approx \frac{1}{R}$.

For the magnetic field,

$$\vec{B} = \vec{\nabla}\times\vec{A} \tag{3.29}$$

$$\vec{B} = \vec{\nabla}\times\left(\frac{\vec{v}}{c^2}\phi\right) \tag{3.30}$$

$$\vec{B} = \frac{1}{c^2}\left[\phi(\vec{\nabla}\times\vec{v}) - \vec{v}\times(\vec{\nabla}\phi)\right] \tag{3.31}$$

| recall that $\vec{\nabla}\times\vec{v} = -\vec{a}\times\vec{\nabla}t_r = \dfrac{\vec{a}\times\vec{R}}{Rc - \vec{R}\cdot\vec{v}}$

| recall that $\vec{\nabla}\phi = \dfrac{qc}{4\pi\epsilon_0}\dfrac{1}{(Rc - \vec{R}\cdot\vec{v})^3}\left[(Rc - \vec{R}\cdot\vec{v})\vec{v} - (c^2 - v^2 + \vec{R}\cdot\vec{a})\vec{R}\right]$

$$\text{| then } \vec{v} \times \vec{v} = 0 \text{ and } \vec{v} \times \vec{\nabla}\phi = \frac{qc}{4\pi\epsilon_0} \frac{1}{(Rc - \vec{R}\cdot\vec{v})^3} \left[-(c^2 - v^2 + \vec{R}\cdot\vec{a})(\vec{v} \times \vec{R}) \right]$$

$$\vec{B} = \frac{1}{c}\frac{q}{4\pi\epsilon_0} \frac{1}{(Rc - \vec{R}\cdot\vec{v})^3} \left[(Rc - \vec{R}\cdot\vec{v})^2 \frac{\vec{a} \times \vec{R}}{Rc - \vec{R}\cdot\vec{v}} + (c^2 - v^2 + \vec{R}\cdot\vec{a})(\vec{v} \times \vec{R}) \right] \tag{3.32}$$

$$\vec{B} = -\frac{1}{c}\frac{q}{4\pi\epsilon_0} \frac{1}{(Rc - \vec{R}\cdot\vec{v})^3} \vec{R} \times \left[\vec{a}(Rc - \vec{R}\cdot\vec{v}) + \vec{v}(c^2 - v^2 + \vec{R}\cdot\vec{a}) \right] \tag{3.33}$$

$$\vec{B} = \frac{1}{c}\frac{q}{4\pi\epsilon_0} \frac{R}{(Rc - \vec{R}\cdot\vec{v})^3} \hat{R} \times \left[(c^2 - v^2)(-\vec{v}) - (\vec{R}\cdot\vec{a})\vec{v} - Rc\vec{a} + (\vec{R}\cdot\vec{v})\vec{a} \right] \tag{3.34}$$

| insert $\hat{R}c$ into first term since $\hat{R} \times \hat{R}c = 0$ anyway

| insert $(\vec{R}\cdot\vec{a})\hat{R}c$ since $\hat{R} \times \vec{R} = 0$ anyway

$$\vec{B} = \frac{1}{c}\frac{q}{4\pi\epsilon_0} \frac{R}{(Rc - \vec{R}\cdot\vec{v})^3} \hat{R} \times \Big[(c^2 - v^2)(\hat{R}c - \vec{v}) + (\vec{R}\cdot\vec{a})\hat{R}c$$
$$-(\vec{R}\cdot\vec{a})\vec{v} - Rc\vec{a} + (\vec{R}\cdot\vec{v})\vec{a} \Big] \tag{3.35}$$

| we can check backwards that the last 4 terms become $\vec{R} \times ((\hat{R}c - \vec{v}) \times \vec{a})$

$$\boxed{\vec{B}(\vec{r}, t) = \frac{1}{c}\frac{q}{4\pi\epsilon_0} \frac{R}{(Rc - \vec{R}\cdot\vec{v})^3} \hat{R} \times \left[(c^2 - v^2)(\hat{R}c - \vec{v}) + \vec{R} \times ((\hat{R}c - \vec{v}) \times \vec{a}) \right]} \tag{3.36}$$

- Term $\frac{1}{c}\frac{q}{4\pi\epsilon_0} \frac{R}{(Rc - \vec{R}\cdot\vec{v})^3} \hat{R} \times (c^2 - v^2)(\hat{R}c - \vec{v})$ is called the velocity field term and it falls as $\approx \frac{1}{R^2}$.

- Term $\frac{1}{c}\frac{q}{4\pi\epsilon_0} \frac{R}{(Rc - \vec{R}\cdot\vec{v})^3} \hat{R} \times \left[\vec{R} \times ((\hat{R}c - \vec{v}) \times \vec{a}) \right]$ is called the acceleration field or the radiation field and it falls as $\approx \frac{1}{R}$.

A quick comparison between the 2 fields give

$$\vec{B}(\vec{r}, t) = \frac{1}{c}\hat{R} \times \vec{E}(\vec{r}, t) \tag{3.37}$$

Thus the magnetic field of a point charge is always perpendicular to the electric field and to the vector $\hat{R}$ that points from the retarded position to the point of observation.

3.3 Example: Point charge in constant velocity

Before we start the example of a point charge with constant velocity, we quickly talk about a point charge that is stationary. This means we set $\vec{a} = 0$ and $\vec{v} = 0$. For the potentials,

$$\phi(\vec{r}) = \frac{qc}{4\pi\epsilon_0} \frac{1}{R(t_r)c} \tag{3.38}$$

| where $R(t_r)$ is simply R

$$\phi(\vec{r}) = \frac{q}{4\pi\epsilon_0 R} \quad \text{(as expected)} \tag{3.39}$$

$$\vec{A}(\vec{r}) = \frac{\vec{v}(t_r)}{c^2}\phi \stackrel{\text{set } \vec{v} = 0}{=} 0 \quad \text{(as expected)} \tag{3.40}$$

Standard Application 1: Point Charge

For the fields, after setting $\vec{a} = 0$ and $\dot{\vec{v}} = 0$,

$$\vec{E}(\vec{r}) = \frac{q}{4\pi\epsilon_0} \frac{R}{(Rc)^3} c^2 \hat{R} c = \frac{q}{4\pi\epsilon_0 R^2} \hat{R} \tag{3.41}$$

which is Coulomb's law as expected. Then,

$$\vec{B}(\vec{r}) = \frac{1}{c}\hat{R} \times \vec{E} = \frac{1}{c}\hat{R} \times \frac{q}{4\pi\epsilon_0 R^2} \hat{R} = 0 \text{ (as expected)} \tag{3.42}$$

For the point charge in constant velocity, the trajectory where the particle passes through the origin at $t = 0$ is $\vec{r}_0(t) = \vec{v}t$. So for this case, we can make the implicit dependence of t_r to be explicit.[2]

$$t_r = t - \frac{|\vec{r} - \vec{r}_0(t_r)|}{c} \tag{3.43}$$

$$|\vec{r} - \vec{v}t_r|^2 = c^2(t - t_r)^2 \tag{3.44}$$

$$r^2 - 2\vec{r}\cdot\vec{v}t_r + v^2 t_r^2 = c^2(t^2 - 2tt_r + t_r^2) \tag{3.45}$$

$$t_r = \frac{(c^2 t - \vec{r}\cdot\vec{v}) \pm \sqrt{(c^2 t - \vec{r}\cdot\vec{v})^2 - (c^2 - v^2)(c^2 t^2 - r^2)}}{(c^2 - v^2)} \tag{3.46}$$

| to fix the sign, we set $\vec{v} = 0$ and require $t_r < t$
| so the negative sign is the meaningful one

$$t_r = \frac{(c^2 t - \vec{r}\cdot\vec{v}) - \sqrt{(c^2 t - \vec{r}\cdot\vec{v})^2 - (c^2 - v^2)(c^2 t^2 - r^2)}}{(c^2 - v^2)} \tag{3.47}$$

In the Lienard–Wiechert potentials, we have the common denominator (C.D.),

$$\text{C.D.} = R(t_r)c - \vec{R}(t_r)\cdot\vec{v}(t_r) \tag{3.48}$$

| recall that $R(t_r) = |\vec{r} - \vec{r}_0(t_r)| = |\vec{r} - \vec{v}t_r| = c(t - t_r)$

$$\text{C.D.} = c^2(t - t_r) - (\vec{r} - \vec{v}t_r)\cdot\vec{v} \tag{3.49}$$

$$\text{C.D.} = c^2(t - t_r) - \vec{r}\cdot\vec{v} + v^2 t_r \tag{3.50}$$

$$\text{C.D.} = (v^2 - c^2)t_r + c^2 t - \vec{r}\cdot\vec{v} \tag{3.51}$$

| from above solution of t_r write $-((c^2 - v^2)t_r - (c^2 t - \vec{r}\cdot\vec{v})) = \sqrt{\cdots}$

$$\text{C.D.} = \sqrt{(c^2 t - \vec{r}\cdot\vec{v})^2 - (c^2 - v^2)(c^2 t^2 - r^2)} \tag{3.52}$$

| further rewrite: define $\vec{r}_p = \vec{r} - \vec{v}t$ where $\vec{r}_0(t) = \vec{v}t$ is the present position

$$\text{C.D.} = \left[c^4 t^2 + (\vec{r}\cdot\vec{v})^2 - 2c^2 t(\vec{r}\cdot\vec{v}) + r^2 c^2 - c^4 t^2 - r^2 v^2 + c^2 v^2 t^2\right]^{1/2} \tag{3.53}$$

$$\text{C.D.} = \sqrt{r^2 v^2 \cos^2\theta - r^2 v^2 + c^2 r_p^2} \tag{3.54}$$

$$\text{C.D.} = c r_p \sqrt{1 - \frac{r^2 v^2}{c^2 r_p^2} \sin^2\theta} \tag{3.55}$$

| use geometry as in figure 3.2

$$\text{C.D.} = c r_p \sqrt{1 - \frac{v^2}{c^2} \sin^2\alpha} \tag{3.56}$$

---

[2] This is a quadratic equation. Next level will be cubic, quartic and so on, so making t_r the subject is much harder.

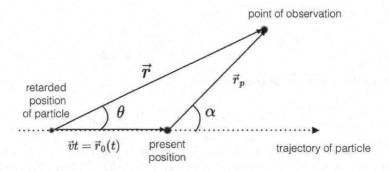

FIGURE 3.2
Somehow the field points from the present position but the information came from the retarded position. We will understand why in Part II. Note the sine rule: $\frac{\sin\theta}{r_p} = \frac{\sin(180^\circ - \alpha)}{r} \implies \frac{\sin^2\theta}{r_p^2} = \frac{\sin^2\alpha}{r^2}$.

The retarded potentials are

$$\phi(\vec{r},t) = \frac{q}{4\pi\epsilon_0 r_p \sqrt{1 - \frac{v^2}{c^2}\sin^2\alpha}} \quad \text{and} \quad \vec{A}(\vec{r},t) = \frac{\vec{v}}{c^2}\phi(\vec{r},t) = \frac{q\vec{v}}{4\pi\epsilon_0 c^2 r_p \sqrt{1 - \frac{v^2}{c^2}\sin^2\alpha}} \quad (3.57)$$

We shall calculate the fields using the explicit formulae for $\vec{E}$ and $\vec{B}$ and start by setting $\vec{a} = 0$ so only the velocity field term survives.

$$\vec{E}(\vec{r},t) = \frac{q}{4\pi\epsilon_0} \frac{R}{(Rc - \vec{R}\cdot\vec{v})^3}(c^2 - v^2)(\hat{R}c - \vec{v}) \quad (3.58)$$

$$\mid \text{recall that } Rc - \vec{R}\cdot\vec{v} = cr_p\sqrt{1 - \frac{v^2}{c^2}\sin^2\alpha}$$

$$\mid \text{note that } R(\hat{R}c - \vec{v}) = \vec{R}c - R\vec{v} = (\vec{r} - \vec{v}t_r)c - c(t - t_r)\vec{v} = (\vec{r} - \vec{v}t)c = c\vec{r}_p$$

$$\vec{E}(\vec{r},t) = \frac{q}{4\pi\epsilon_0} \frac{c\vec{r}_p(c^2 - v^2)}{c^3 r_p^3 \left(1 - \frac{v^2}{c^2}\sin^2\alpha\right)^{3/2}} \quad (3.59)$$

$$\boxed{\vec{E}(\vec{r},t) = \frac{q}{4\pi\epsilon_0} \frac{1 - \frac{v^2}{c^2}}{\left(1 - \frac{v^2}{c^2}\sin^2\alpha\right)^{3/2}} \frac{\hat{r}_p}{r_p^2}} \quad (3.60)$$

So $\vec{E}$ points along the line from the <u>present</u> position of the particle to the point of observation! The fact that the information came from the retarded position and yet the vector points from the present position is due to an extraordinary coincidence which we will uncover in Part II.

The magnetic field is obtained by,

$$\vec{B}(\vec{r},t) = \frac{1}{c}\hat{R} \times \vec{E}(\vec{r},t) \quad (3.61)$$

$$\mid \text{rewrite } \hat{R} = \frac{\vec{R}}{R} = \frac{\vec{r} - \vec{v}t_r}{R} = \frac{\vec{r}_p + \vec{v}t - \vec{v}t_r}{c(t - t_r)} = \frac{\vec{r}_p}{R} + \frac{\vec{v}}{c}$$

$$\mid \text{since } \vec{E} \text{ is in the } \vec{r}_p \text{ direction, } \vec{r}_p \times \vec{E} = 0$$

$$\boxed{\vec{B}(\vec{r},t) = \frac{1}{c^2}\left(\vec{v} \times \vec{E}(\vec{r},t)\right)} \quad (3.62)$$

Standard Application 1: Point Charge

Thus the B-field is perpendicular to both $\vec{v}$ and $\vec{E}$, so the B-field forms circles around the trajectory of the charge.

For a slow-moving charge, where $v \ll c$ or $v^2 \ll c^2$, take $\frac{v^2}{c^2} \to 0$,

$$\vec{E}(\vec{r},t) \approx \frac{q}{4\pi\epsilon_0} \frac{\hat{r}_p}{r_p^2} \quad \text{(which is Coulomb's law)} \tag{3.63}$$

$$\vec{B}(\vec{r},t) \approx \frac{1}{c^2}\vec{v} \times \frac{q}{4\pi\epsilon_0} \frac{\hat{r}_p}{r_p^2} \tag{3.64}$$

$$\bigg| \quad \text{recall that } \frac{1}{c^2} = \epsilon_0\mu_0$$

$$\vec{B}(\vec{r},t) = \frac{\mu_0 q}{4\pi} \frac{\vec{v} \times \hat{r}_p}{r_p^2} \tag{3.65}$$

which is Biot-Savart's law. Thus, Coulomb's law and Biot-Savart's law are only applicable for a slow-moving point charge.

4

Standard Application 2: Radiation

Note that in the example of a point charge with constant velocity, the solutions for the fields are not waves. We suspect that the reason is because we set $\vec{a} = 0$ in that example. Thus, we can deduce that accelerating charges and changing currents are needed to generate electromagnetic waves/radiation.

Waves carry energy from the source and the energy never comes back. For EM waves, we can calculate the radiated power using the Poynting vector since it is the energy density current:

$$P_{\text{rad}} = \lim_{r \to \infty} P(r) = \lim_{r \to \infty} \oint \vec{S} \cdot d\vec{A} \tag{4.1}$$

$$= \lim_{r \to \infty} \frac{1}{\mu_0} \oint (\vec{E} \times \vec{B}) \cdot d\vec{A} \tag{4.2}$$

So if $E \sim \frac{1}{r^2}$ (or faster) and if $B \sim \frac{1}{r^2}$ (or faster) then $S \sim \frac{1}{r^4}$ (or faster) and $SdA \sim \frac{1}{r^4}r^2$ so $P_{\text{rad}} \xrightarrow{r \to \infty} 0$. This means that the velocity field terms do not give rise to radiation. Thus radiation involves acceleration field (or radiation field) terms where $B, E \sim \frac{1}{r}$, so that $SdA \sim \frac{1}{r^2}r^2$ and $P_{\text{rad}} \xrightarrow{r \to \infty}$ finite.

4.1 From point charge

Recall that the electric field of a point charge in arbitrary motion is

$$\vec{E}(\vec{r},t) = \frac{q}{4\pi\epsilon_0} \frac{R}{(Rc - \vec{R} \cdot \vec{v})^3} \left[(c^2 - v^2)(\hat{R}c - \vec{v}) + \vec{R} \times ((\hat{R}c - \vec{v}) \times \vec{a}) \right] \tag{4.3}$$

$$\text{and,} \quad \vec{B}(\vec{r},t) = \frac{1}{c}\hat{R} \times \vec{E}(\vec{r},t) \tag{4.4}$$

Recall that the first term is the velocity field and the second term is the acceleration/radiation field.

The Poynting vector is

$$\vec{S} = \frac{1}{\mu_0}(\vec{E} \times \vec{B}) \tag{4.5}$$

$$\vec{S} = \frac{1}{\mu_0 c}(\vec{E} \times (\hat{R} \times \vec{E})) \tag{4.6}$$

| use product rule: $\vec{A} \times (\vec{B} \times \vec{C}) = \vec{B}(\vec{A} \cdot \vec{C}) - \vec{C}(\vec{A} \cdot \vec{B})$

$$\vec{S} = \frac{1}{\mu_0 c} \left(\vec{E}^2 \hat{R} - (\hat{R} \cdot \vec{E})\vec{E} \right) \tag{4.7}$$

| then note that $\hat{R} \cdot (\vec{R} \times ((\hat{R}c - \vec{v}) \times \vec{a})) = 0$

DOI: 10.1201/9781003515210-4

Standard Application 2: Radiation

| and $\hat{R} \cdot$ "velocity field terms" $\sim \frac{1}{R^2}$ does not cause radiation
| and so drop all velocity field terms

$$\vec{S} \approx \frac{1}{\mu_0 c} \vec{E}^2 \hat{R} \tag{4.8}$$

$$\vec{S} \approx \frac{1}{\mu_0 c} \left(\frac{q}{4\pi\epsilon_0} \frac{R}{(Rc - \vec{R} \cdot \vec{v})^3} \right)^2 \left[\vec{R} \times ((\hat{R}c - \vec{v}) \times \vec{a}) \right] \cdot \left[\vec{R} \times ((\hat{R}c - \vec{v}) \times \vec{a}) \right] \hat{R} \tag{4.9}$$

To make the calculations easier, we carry on with the calculations in a frame where the charge is instantaneously at rest at time t_r, so set $\vec{v} = 0$ but $\vec{a} \neq 0$. We will generalise this in section 8.6.

$$\vec{S} = \frac{1}{\mu_0 c} \left(\frac{q}{4\pi\epsilon_0} \frac{R}{R^3 c^3} \right)^2 \left(\vec{R} \times (\hat{R}c \times \vec{a}) \right) \cdot \left(\vec{R} \times (\hat{R}c \times \vec{a}) \right) \hat{R} \tag{4.10}$$

| use product rule: $\vec{A} \times (\vec{B} \times \vec{C}) = \vec{B}(\vec{A} \cdot \vec{C}) - \vec{C}(\vec{A} \cdot \vec{B})$
| so, $\vec{R} \times (\hat{R} \times \vec{a}) = \hat{R}(\vec{R} \cdot \vec{a}) - \vec{a}(\vec{R} \cdot \hat{R}) = \hat{R}(\vec{R} \cdot \vec{a}) - R\vec{a}$

$$\vec{S} = \frac{1}{\mu_0 c} \left(\frac{q}{4\pi\epsilon_0} \right)^2 \frac{1}{R^4 c^6} c^2 \left(\hat{R}(\vec{R} \cdot \vec{a}) - R\vec{a} \right) \cdot \left(\hat{R}(\vec{R} \cdot \vec{a}) - R\vec{a} \right) \hat{R} \tag{4.11}$$

| recall $\epsilon_0 = \frac{1}{\mu_0 c^2}$ and carry out the dot product

$$\vec{S} = \frac{1}{\mu_0 c} \left(\frac{\mu_0 q}{4\pi R} \right)^2 \frac{1}{R^2} \left(R^2 a^2 - R^2 (\hat{R} \cdot \vec{a})^2 \right) \hat{R} \tag{4.12}$$

$$\vec{S} = \frac{1}{\mu_0 c} \left(\frac{\mu_0 q}{4\pi R} \right)^2 \left(a^2 - (\hat{R} \cdot \vec{a})^2 \right) \hat{R} \tag{4.13}$$

| write $\hat{R} \cdot \vec{a} = a \cos \theta$

$$\vec{S} = \frac{1}{\mu_0 c} \left(\frac{\mu_0 q}{4\pi R} \right)^2 \left(a^2 - a^2 \cos^2 \theta \right) \hat{R} \tag{4.14}$$

$$\boxed{\vec{S} = \frac{\mu_0 q^2 a^2}{16\pi^2 c} \frac{\sin^2 \theta}{R^2} \hat{R}} \tag{4.15}$$

Thus, power is emitted like a donut about the direction of instantaneous acceleration. No radiation is emitted in the forward or backward direction.

The power radiated into a patch of area $R^2 \sin\theta d\theta d\phi = R^2 d\Omega$ is [1]

$$dP = \vec{S} \cdot d\vec{A} = |\vec{S}| R^2 d\Omega \tag{4.16}$$

$$\frac{dP}{d\Omega} = \frac{\mu_0 q^2 a^2}{16\pi^2 c} \sin^2 \theta \tag{4.17}$$

The total power emitted is found by integrating over the entire solid angle.

$$P = \int dP \tag{4.18}$$

$$P = \int \frac{dP}{d\Omega} d\Omega \quad \text{or,} \quad \int \vec{S} \cdot d\vec{A} \tag{4.19}$$

---
[1]Reminder: solid angle is defined as $d\Omega = \sin\theta d\theta d\phi$

$$P = \frac{\mu_0 q^2 a^2}{16\pi^2 c} \int_0^{2\pi} d\phi \int_0^{\pi} d\theta \sin^2\theta \sin\theta \qquad (4.20)$$

$$|\text{ note that } \int_0^{2\pi} d\phi = 2\pi \text{ and } \int_0^{\pi} \sin^3\theta d\theta = \frac{4}{3}$$

$$\boxed{P = \frac{\mu_0 q^2 a^2}{6\pi c}} \qquad (4.21)$$

This is Larmor's formula for slow moving, accelerating point charge. We shall see the Lienard's generalised version of Larmor's formula (for point charge) for any velocity in section 8.6.

4.2 From hertzian dipole

Consider an electric dipole driven as shown in figure 4.1.

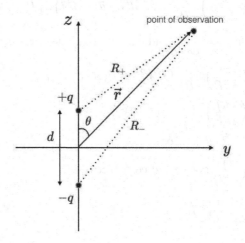

FIGURE 4.1
The setup for the Hertzian dipole radiation source.

We need to calculate the retarded potentials, then get the fields and finally get the power radiated.

The retarded potentials are,

$$\phi(\vec{r}, t) = \frac{1}{4\pi\epsilon_0} \int \frac{\rho(\vec{r}', t_r)}{|\vec{r} - \vec{r}'|} dV' \qquad (4.22)$$

$$| \text{ with } \rho(\vec{r}', t_r) = q_0 \cos(\omega t_r^+) \delta^3(\vec{r}' - \frac{d}{2}\hat{z}) + (-q_0)\cos(\omega t_r^-)\delta^3(\vec{r}' + \frac{d}{2}\hat{z})$$

$$| \text{ where } t_r^+ = t - \frac{R_+}{c}, \; R_+ = \left|\vec{r} - \frac{d}{2}\hat{z}\right| \text{ and } t_r^- = t - \frac{R_-}{c}, \; R_- = \left|\vec{r} + \frac{d}{2}\hat{z}\right|$$

$$\phi(\vec{r}, t) = \frac{1}{4\pi\epsilon_0}\left(\frac{q_0 \cos\left(\omega\left(t - \frac{R_+}{c}\right)\right)}{R_+} - \frac{q_0 \cos\left(\omega\left(t - \frac{R_-}{c}\right)\right)}{R_-}\right) \qquad (4.23)$$

Standard Application 2: Radiation

| cosine rule: $R_\pm^2 = r^2 + \left(\frac{d}{2}\right)^2 - 2r\left(\frac{d}{2}\right)\cos\left(180^0 - \theta\atop\theta\right)$

| $\phantom{\text{cosine rule: }R_\pm^2} = r^2 + \left(\frac{d}{2}\right)^2 \mp 2r\left(\frac{d}{2}\right)\cos\theta$

| Approximation 1: physical dipole $\to$ perfect dipole $\Rightarrow d \ll r$

| so $R_\pm = \sqrt{r^2\left(1 + \left(\frac{d}{2r}\right)^2 \mp \frac{d}{r}\cos\theta\right)} \approx r\sqrt{1 \mp \frac{d}{r}\cos\theta} \approx r\left(1 \mp \frac{1}{2}\frac{d}{r}\cos\theta\right)$

| so $\frac{1}{R_\pm} \approx \frac{1}{r}\left(1 \mp \left(-\frac{1}{2}\right)\frac{d}{r}\cos\theta\right) = \frac{1}{r}\left(1 \pm \frac{d}{2r}\cos\theta\right)$

| so $\cos\left(\omega\left(t - \frac{R_\pm}{c}\right)\right) \approx \cos\left(\omega\left(t - \frac{r}{c}\left(1 \mp \frac{d}{2r}\cos\theta\right)\right)\right)$

$\phantom{\text{so }\cos\left(\omega\left(t - \frac{R_\pm}{c}\right)\right)} = \cos\left(\omega\left(t - \frac{r}{c}\right) \pm \frac{\omega d}{2c}\cos\theta\right)$

| then use cosine addition formula $\cos(A \pm B) = \cos A \cos B \mp \sin A \sin B$
| Approximation 2: perfect dipole:

| $\Rightarrow d \ll \lambda \Rightarrow d \ll \frac{c}{f} \Rightarrow d \ll \frac{2\pi c}{\omega} \Rightarrow d \ll \frac{c}{\omega} \Rightarrow \frac{\omega d}{c} \ll 1$

| so $\cos\left(\frac{\omega d}{2c}\cos\theta\right) \approx 1$ and $\sin\left(\frac{\omega d}{2c}\cos\theta\right) \approx \frac{\omega d}{2c}\cos\theta$

$\phi(r, \theta, t) = \frac{q_0}{4\pi\epsilon_0}\left\{\left[\cos\left(\omega\left(t - \frac{r}{c}\right)\right) - \frac{\omega d}{2c}\cos\theta \sin\left(\omega\left(t - \frac{r}{c}\right)\right)\right]\frac{1}{r}\left(1 + \frac{d}{2r}\cos\theta\right)\right.$
$\left. - \left[\cos\left(\omega\left(t - \frac{r}{c}\right)\right) + \frac{\omega d}{2c}\cos\theta \sin\left(\omega\left(t - \frac{r}{c}\right)\right)\right]\frac{1}{r}\left(1 - \frac{d}{2r}\cos\theta\right)\right\}$ (4.24)

| expand and 4 terms cancel pairwise, 4 terms add pairwise
| denote $p_0 = q_0 d$ as the dipole moment

$= \frac{p_0 \cos\theta}{4\pi\epsilon_0 r}\left\{-\frac{\omega}{c}\sin\left(\omega\left(t - \frac{r}{c}\right)\right) + \frac{1}{r}\cos\left(\omega\left(t - \frac{r}{c}\right)\right)\right\}$ (4.25)

| Approximation 3: far field or radiation zone: $r \gg \lambda \Rightarrow r \gg \frac{c}{\omega} \Rightarrow \frac{1}{r} \ll \frac{\omega}{c}$

| so we drop the second term

$\phi(\vec{r}, t) = -\frac{p_0 \omega}{4\pi\epsilon_0 c}\frac{\cos\theta}{r}\sin\left(\omega\left(t - \frac{r}{c}\right)\right)$ (4.26)

Now for $\vec{A}(\vec{r}, t)$,

$\vec{A}(\vec{r}, t) = \frac{\mu_0}{4\pi}\int_{-d/2}^{d/2}\frac{\vec{J}(\vec{r}', t_r)}{R}dz$ (4.27)

$\vec{A}(\vec{r}, t) = \frac{\mu_0}{4\pi}\int_{-d/2}^{d/2}\frac{\vec{J}\left(\vec{r}', t - \frac{R}{c}\right)}{R}dz$ (4.28)

| take zeroth order approximation: $R \approx r$. Justification right below.

$\vec{A}(\vec{r}, t) \approx \frac{\mu_0}{4\pi r}\int_{-d/2}^{d/2}\vec{J}\left(\vec{r}', t - \frac{r}{c}\right)dz$ (4.29)

| use identity: $\int \vec{J} dV = \frac{d\vec{p}}{dt}$ with proof in footnote

$$\vec{A}(\vec{r},t) = \frac{\mu_0}{4\pi r}\frac{d\vec{p}}{dt} \qquad (4.30)$$

$$\vec{A}(\vec{r},t) = \frac{\mu_0}{4\pi r}\frac{d}{dt}p_0 \cos\left(\omega\left(t-\frac{r}{c}\right)\right)\hat{z} \qquad (4.31)$$

$$\vec{A}(r,\theta,t) = -\frac{\mu_0 p_0 \omega}{4\pi r}\sin\left(\omega\left(t-\frac{r}{c}\right)\right)\hat{z} \qquad (4.32)$$

Note[2] that $\vec{A}(r,\theta,t)$ is not subjected to Approximation 3. Now we justify the earlier approximation: $R \approx r$. Notice that $\frac{d\vec{p}}{dt}$ brings in $p_0 = q_0 d$ and if we approximate $\frac{1}{R} \approx \frac{1}{r}\left(1 + \frac{d}{2r}\cos\theta\right)$, we will have d^2 terms which are too small.

The fields are,

$$\vec{E} = -\vec{\nabla}\phi - \frac{\partial \vec{A}}{\partial t} \qquad (4.36)$$

$$\vec{E} = -\frac{\partial \phi}{\partial r}\hat{r} - \frac{1}{r}\frac{\partial \phi}{\partial \theta}\hat{\theta} - \frac{\partial \vec{A}}{\partial t} \qquad (4.37)$$

| note that $\frac{1}{r}\frac{\partial \phi}{\partial \theta}\hat{\theta}$ gives $\frac{1}{r^2}$ term which is not radiative, so drop it

$$\vec{E} \approx -\frac{\partial \phi}{\partial r}\hat{r} - \frac{\partial \vec{A}}{\partial t} \qquad (4.38)$$

$$\vec{E} = \frac{p_0 \omega \cos\theta}{4\pi\epsilon_0 c}\left[\frac{\sin\left(\omega\left(t-\frac{r}{c}\right)\right)}{r^2} - \frac{\omega}{rc}\cos\left(\omega\left(t-\frac{r}{c}\right)\right)\right]\hat{r} + \frac{\mu_0 p_0 \omega^2}{4\pi r}\cos\left(\omega\left(t-\frac{r}{c}\right)\right)\hat{z} \qquad (4.39)$$

| recall that $\hat{z} = \cos\theta\hat{r} - \sin\theta\hat{\theta}$ and $c^2 = \frac{1}{\mu_0 \epsilon_0}$

| so $\cos\theta\cos\left(\omega\left(t-\frac{r}{c}\right)\right)$ from $\frac{\partial \vec{A}}{\partial t}$ cancels that from $\frac{\partial \phi}{\partial r}$

| drop $\cos\theta\frac{\sin\left(\omega\left(t-\frac{r}{c}\right)\right)}{r^2}$ term as it is $\sim \frac{1}{r^2}$

$$\boxed{\vec{E} \approx -\frac{\mu_0 p_0 \omega^2}{4\pi}\frac{\sin\theta}{r}\cos\left(\omega\left(t-\frac{r}{c}\right)\right)\hat{\theta}} \qquad (4.40)$$

---

[2]Proof:

$$\frac{d\vec{p}}{dt} = \frac{d}{dt}\int \rho\vec{r}dV = \int \frac{\partial \rho}{\partial t}\vec{r}dV \qquad (4.33)$$

| use charge continuity equation: $\frac{\partial \rho}{\partial t} = -\vec{\nabla}\cdot\vec{J}$

$$\frac{d\vec{p}}{dt} = -\int \left(\vec{\nabla}\cdot\vec{J}\right)\vec{r}dV \qquad (4.34)$$

| use product rule: $\vec{\nabla}\cdot(\vec{J}f) = (\vec{\nabla}\cdot\vec{J})f + \vec{J}\cdot(\vec{\nabla}f)$

| so $(\vec{\nabla}\cdot\vec{J})x = \vec{\nabla}\cdot(x\vec{J}) - \vec{J}\cdot(\vec{\nabla}x) = \vec{\nabla}\cdot(x\vec{J}) - J_x$

| so $(\vec{\nabla}\cdot\vec{J})\vec{r} = \vec{\nabla}\cdot(x\vec{J})\hat{x} + \vec{\nabla}\cdot(y\vec{J})\hat{y} + \vec{\nabla}\cdot(z\vec{J})\hat{z} - \vec{J}$

| and first 3 terms: $\int \vec{\nabla}\cdot(x\vec{J})dV\hat{x}$ are zero when we use divergence theorem

| and current $\vec{J}$ is inside a sufficiently large volume

$$\frac{d\vec{p}}{dt} = \int \vec{J}dV \quad \text{(proved)} \qquad (4.35)$$

Standard Application 2: Radiation

which is a spherical wave expression.

$$\vec{B} = \vec{\nabla} \times \vec{A} \tag{4.41}$$

$$\vec{B} = \frac{1}{r}\left[\frac{\partial}{\partial r}(rA_\theta) - \frac{\partial A_r}{\partial \theta}\right]\hat{\phi} \tag{4.42}$$

| note that $\frac{1}{r}\frac{\partial A_r}{\partial \theta}$ gives $\frac{1}{r^2}$ term which is not radiative, so drop it

$$\vec{B} = \frac{1}{r}\frac{\partial}{\partial r}\left(r\left(\frac{\mu_0 p_0 \omega}{4\pi r}\sin\theta \sin\left(\omega\left(t - \frac{r}{c}\right)\right)\right)\right)\hat{\phi} \tag{4.43}$$

$$\boxed{\vec{B} = -\frac{\mu_0 p_0 \omega^2}{4\pi c}\frac{\sin\theta}{r}\cos\left(\omega\left(t - \frac{r}{c}\right)\right)\hat{\phi}} \tag{4.44}$$

again, a spherical wave expansion. The fields have the following features:

- $\vec{E}$ and $\vec{B}$ are in phase and they oscillate with (angular) frequency ω.

- The waves are travelling in the $\hat{r}$ direction while $\vec{E}$ and $\vec{B}$ are in the $\hat{\theta}$ and $\hat{\phi}$ directions, respectively. This means $\vec{E}$ and $\vec{B}$ are transverse and mutually perpendicular.

- The amplitudes are in the ratio $\frac{|\vec{E}|}{|\vec{B}|} = c$ which we already know from the vacuum solutions of electrodynamics.

Finally, we can proceed to calculate the Poynting vector.

$$\vec{S} = \frac{1}{\mu_0}(\vec{E} \times \vec{B}) \tag{4.45}$$

| note that $\hat{\theta} \times \hat{\phi} = \hat{r}$

$$\vec{S} = \frac{\mu_0}{c}\left[\frac{p_0\omega^2}{4\pi}\left(\frac{\sin\theta}{r}\right)\cos\left(\omega\left(t - \frac{r}{c}\right)\right)\right]^2 \hat{r} \tag{4.46}$$

We are usually more interested in the time-averaged power than the instantaneous power,

$$\langle\vec{S}\rangle_\text{time} = \frac{1}{T}\int_0^T \vec{S}\,dt \tag{4.47}$$

$$\langle\vec{S}\rangle_\text{time} = \frac{\omega}{2\pi}\int_0^{2\pi/\omega} \frac{\mu_0}{c}\left(\frac{p_0\omega^2}{4\pi}\frac{\sin\theta}{r}\right)^2 \cos^2\left(\omega\left(t - \frac{r}{c}\right)\right)dt\,\hat{r} \tag{4.48}$$

| averaging $\cos^2$ over a period gives $\frac{1}{2}$

$$\boxed{\langle\vec{S}\rangle_\text{time} = \frac{\mu_0 p_0^2 \omega^4}{32\pi^2 c}\frac{\sin^2\theta}{r^2}\hat{r}} \tag{4.49}$$

Power radiated into a patch of area $r^2 \sin\theta\,d\theta\,d\phi = r^2 d\Omega$ is

$$dP = \langle\vec{S}\rangle_\text{time} \cdot d\vec{A} = \left|\langle\vec{S}\rangle_\text{time}\right|r^2 d\Omega = \frac{\mu_0 p_0^2 \omega^4}{32\pi^2 c}\sin^2\theta\,d\Omega \tag{4.50}$$

$$\text{Total Power} = P = \int dP \tag{4.51}$$

$$P = \int \frac{dP}{d\Omega}d\Omega \tag{4.52}$$

$$P = \frac{\mu_0 p_0^2 \omega^4}{32\pi^2 c} \int_0^{2\pi} d\phi \int_0^{\pi} \sin^3\theta d\theta \qquad (4.53)$$

| recall that $\int_0^{2\pi} d\phi = 2\pi$ and $\int_0^{\pi} \sin^3\theta d\theta = \frac{4}{3}$

$$\boxed{P = \frac{\mu_0 p_0^2 \omega^4}{12\pi c}} \qquad (4.54)$$

This is Larmor's formula for power radiated by a Hertzian dipole.

4.3 From arbitrary distribution

We consider, in general, some configuration of charge and current localised in some finite volume near the origin.

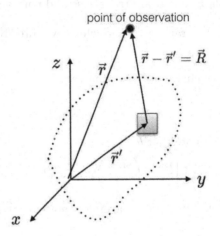

FIGURE 4.2
The setup for an arbitrary distribution near the origin.

Again we start with the retarded potentials.

$$\phi(\vec{r}, t) = \frac{1}{4\pi\epsilon_0} \int \frac{\rho(\vec{r}', t_r)}{|\vec{r} - \vec{r}'|} dV' \qquad (4.55)$$

| where cosine rule gives $|\vec{r} - \vec{r}'| = R = \sqrt{r^2 + r'^2 - 2\vec{r}\cdot\vec{r}'}$
| Approximation 1:
| point of observation is much further away than source dimension: $r' \ll r$
| so, $R = \sqrt{r^2\left(1 + \left(\frac{r'}{r}\right)^2 - 2\frac{\vec{r}\cdot\vec{r}'}{r^2}\right)} \approx r\left(1 - \frac{\vec{r}\cdot\vec{r}'}{r^2}\right)$ so, $\frac{1}{R} \approx \frac{1}{r}\left(1 + \frac{\vec{r}\cdot\vec{r}'}{r^2}\right)$
| then, $\rho(\vec{r}', t_r) = \rho\left(\vec{r}', t - \frac{R}{c}\right) \approx \rho\left(\vec{r}', t - \frac{r}{c} + \frac{\hat{r}\cdot\vec{r}'}{c}\right)$
| $\approx \rho(\vec{r}', \tilde{t}) + \dot\rho(\vec{r}', \tilde{t})\left(\frac{\hat{r}\cdot\vec{r}'}{c}\right) + \cdots$

Standard Application 2: Radiation

| where $\dot{\rho}(\vec{r}',\tilde{t}) = \dfrac{\partial \rho(\vec{r},\tilde{t})}{\partial t}$ and $\tilde{t} = t - \dfrac{r}{c}$
| Approximation 2:
| terms $\dfrac{1}{2}\ddot{\rho}(\vec{r}',\tilde{t})\left(\dfrac{\hat{r}\cdot\vec{r}'}{c}\right)^2 + \cdots$ can be dropped if $r' \ll \dfrac{c}{|\ddot{\rho}/\dot{\rho}|}, \cdots$

$$\phi(\vec{r},t) \approx \frac{1}{4\pi\epsilon_0}\int \frac{1}{r}\left(1+\frac{\vec{r}\cdot\vec{r}'}{r^2}\right)\left(\rho(\vec{r}',\tilde{t})+\dot{\rho}(\vec{r}',\tilde{t})\left(\frac{\hat{r}\cdot\vec{r}'}{c}\right)\right)dV' \quad (4.56)$$

$$\phi(\vec{r},t) \approx \frac{1}{4\pi\epsilon_0 r}\left[\int \rho(\vec{r}',\tilde{t})dV' + \int \frac{\hat{r}\cdot\vec{r}'}{r}\rho(\vec{r}',\tilde{t})dV' + \int \frac{\hat{r}\cdot\vec{r}'}{c}\dot{\rho}(\vec{r}',\tilde{t})dV'\right] \quad (4.57)$$

| so, $\int \rho(\vec{r}',\tilde{t})dV' = Q$ the total charge
| so, $\int \dfrac{\hat{r}\cdot\vec{r}'}{r}\rho(\vec{r}',\tilde{t})dV' = \dfrac{\hat{r}}{r}\cdot\int \vec{r}'\rho(\vec{r}',\tilde{t})dV' = \dfrac{\hat{r}\cdot\vec{p}(\tilde{t})}{r}$
| so, $\int \dfrac{\hat{r}\cdot\vec{r}'}{c}\dot{\rho}(\vec{r}',\tilde{t})dV' = \dfrac{d}{dt}\dfrac{\hat{r}}{c}\cdot\int \vec{r}'\rho(\vec{r}',\tilde{t})dV' = \dfrac{\hat{r}\cdot\dot{\vec{p}}(\tilde{t})}{c}$

$$\phi(\vec{r},t) = \frac{1}{4\pi\epsilon_0}\Bigg(\underbrace{\frac{Q}{r}}_{\text{monopole}}+\underbrace{\frac{\hat{r}\cdot\vec{p}(\tilde{t})}{r^2}}_{\text{dipole}}+\frac{\hat{r}\cdot\dot{\vec{p}}(\tilde{t})}{rc}\Bigg) \quad (4.58)$$

| Approximation 3: the monopole and dipole terms are static
| hence they are not radiative and so drop them

$$\phi(\vec{r},t) \approx \frac{1}{4\pi\epsilon_0}\frac{\hat{r}\cdot\dot{\vec{p}}(\tilde{t})}{rc} \quad (4.59)$$

Now for $\vec{A}(\vec{r},t)$,

$$\vec{A}(\vec{r},t) = \frac{\mu_0}{4\pi}\int \frac{\vec{J}(\vec{r}',t_r)}{R}dV' \quad (4.60)$$

| recall that we take the zeroth order approximation: $R \approx r$
| also recall the identity: $\int \vec{J}\left(\vec{r}',t-\dfrac{r}{c}\right)dV' = \dfrac{d\vec{p}}{dt} = \dot{\vec{p}}(\tilde{t})$

$$\vec{A}(\vec{r},t) \approx \frac{\mu_0}{4\pi}\frac{\dot{\vec{p}}(\tilde{t})}{r} \quad (4.61)$$

We can proceed to calculate the fields,

$$\vec{E}(\vec{r},t) = -\vec{\nabla}\phi - \frac{\partial \vec{A}}{\partial t} \quad (4.62)$$

| in $\vec{\nabla}\phi$, after product rule, we drop the $\vec{\nabla}\dfrac{1}{r}$ term as it gives $\dfrac{1}{r^2}$ term

$$\vec{E}(\vec{r},t) \approx -\frac{1}{4\pi\epsilon_0}\frac{\vec{\nabla}\left(\hat{r}\cdot\dot{\vec{p}}(\tilde{t})\right)}{rc} - \frac{\mu_0}{4\pi}\frac{\ddot{\vec{p}}(\tilde{t})}{r} \quad (4.63)$$

| so $\vec{\nabla}\left(\hat{r}\cdot\dot{\vec{p}}(\tilde{t})\right) = \vec{\nabla}\dot{p}_r = \dfrac{\partial \dot{p}_r}{\partial r}\hat{r} = \dfrac{\partial \dot{p}_r}{\partial \tilde{t}}\dfrac{\partial \tilde{t}}{\partial r}\hat{r} = \ddot{p}_r\left(-\dfrac{1}{c}\right)\hat{r} = -\dfrac{1}{c}\left(\hat{r}\cdot\ddot{\vec{p}}(\tilde{t})\right)\hat{r}$

$$\vec{E}(\vec{r},t) = \frac{1}{4\pi\epsilon_0 c^2}\frac{\hat{r}\cdot\ddot{\vec{p}}(\tilde{t})}{r}\hat{r} - \frac{\mu_0}{4\pi}\frac{\ddot{\vec{p}}(\tilde{t})}{r} \quad (4.64)$$

| recall that $\dfrac{1}{c^2} = \epsilon_0 \mu_0$

$$\vec{E}(\vec{r},t) = \frac{\mu_0}{4\pi r}\left((\hat{r}\cdot\ddot{\vec{p}}(\tilde{t}))\hat{r} - \ddot{\vec{p}}(\tilde{t})\right) \tag{4.65}$$

| use product rule: $\vec{A}\times(\vec{B}\times\vec{C}) = \vec{B}(\vec{A}\cdot\vec{C}) - \vec{C}(\vec{A}\cdot\vec{B})$
| so that $\hat{r}\times(\hat{r}\times\ddot{\vec{p}}(\tilde{t})) = \hat{r}(\hat{r}\cdot\ddot{\vec{p}}(\tilde{t})) - \ddot{\vec{p}}(\tilde{t})(\hat{r}\cdot\hat{r})$

$$\vec{E}(\vec{r},t) = \frac{\mu_0}{4\pi r}\left(\hat{r}\times(\hat{r}\times\ddot{\vec{p}}(\tilde{t}))\right) \tag{4.66}$$

$$\vec{B}(\vec{r},t) = \vec{\nabla}\times\vec{A} \tag{4.67}$$

| use product rule: $\vec{\nabla}\times(f\vec{A}) = f(\vec{\nabla}\times\vec{A}) - \vec{A}\times(\vec{\nabla}f)$
| but $\vec{\nabla}\frac{1}{r}$ will give $\frac{1}{r^2}$ term and so we drop it

$$\vec{B}(\vec{r},t) \approx \frac{\mu_0}{4\pi r}\vec{\nabla}\times\dot{\vec{p}}(\tilde{t}) \tag{4.68}$$

| use chain rule: $\vec{\nabla}\times\dot{\vec{p}}(\tilde{t}) = (\vec{\nabla}\tilde{t})\times\ddot{\vec{p}}(\tilde{t})$ and $\vec{\nabla}\tilde{t} = -\frac{1}{c}\hat{r}$

$$\vec{B}(\vec{r},t) = -\frac{\mu_0}{4\pi rc}\hat{r}\times\ddot{\vec{p}}(\tilde{t}) \tag{4.69}$$

To calculate power, we can choose to align $\ddot{\vec{p}}(\tilde{t})$ along the z-axis,

$$\vec{E}(\vec{r},t) = \vec{E}(r,\theta,t) = \frac{\mu_0}{4\pi r}\left(\hat{r}\times(\hat{r}\times\hat{z})\right)\ddot{p}(\tilde{t}) \tag{4.70}$$

| recall $\hat{z} = \cos\theta\hat{r} - \sin\theta\hat{\theta}$, $\hat{r}\times\hat{\theta} = \hat{\phi}$ and $\hat{r}\times\hat{\phi} = -\hat{\theta}$

$$\vec{E}(r,\theta,t) = \frac{\mu_0\ddot{p}(\tilde{t})}{4\pi}\frac{\sin\theta}{r}\hat{\theta} \tag{4.71}$$

$$\vec{B}(\vec{r},t) = \vec{B}(r,\theta,t) \tag{4.72}$$

| recall that $\hat{r}\times\hat{z} = -\sin\theta\hat{\phi}$

$$\vec{B}(r,\theta,t) = \frac{\mu_0\ddot{p}(\tilde{t})}{4\pi c}\frac{\sin\theta}{r}\hat{\phi} \tag{4.73}$$

The Poynting vector is,

$$\vec{S} = \frac{1}{\mu_0}(\vec{E}\times\vec{B}) = \frac{\mu_0}{c}\left(\frac{\ddot{p}(\tilde{t})\sin\theta}{4\pi}\frac{}{r}\right)^2\hat{\theta}\times\hat{\phi} = \frac{\mu_0\ddot{p}(\tilde{t})^2}{16\pi^2 c}\frac{\sin^2\theta}{r^2}\hat{r} \tag{4.74}$$

$$dP = \vec{S}\cdot d\vec{A} = |\vec{S}|r^2\sin\theta d\theta d\phi = \frac{\mu_0\ddot{p}(\tilde{t})^2}{16\pi^2 c}\sin^3\theta d\Omega \tag{4.75}$$

Total power, $$P = \int\frac{dP}{d\Omega}d\Omega \tag{4.76}$$

$$P = \frac{\mu_0\ddot{p}(\tilde{t})^2}{16\pi^2 c}\int_0^{2\pi}d\phi\int_0^{\pi}\sin^3\theta d\theta \tag{4.77}$$

$$P = \frac{\mu_0\ddot{p}(\tilde{t})^2}{16\pi^2 c}\times 2\pi\times\frac{4}{3} \tag{4.78}$$

$$\boxed{P = \frac{\mu_0\ddot{p}(\tilde{t})^2}{6\pi c}} \tag{4.79}$$

The radiation from an arbitrary distribution is again "donut-shaped". This is because in Approximation 2, we have dropped higher order terms (quadrupole, octupole ...) and we have only kept up to the dipole contribution from the source.

5

Standard Application 3: Scattering of EM waves

Scattering is the process where electromagnetic waves impinge on a charged object (scatterer) and disturbs the charged object. The disturbance accelerates the charged object thereby emitting its own electromagnetic waves. The superposition of the incident wave and the emitted wave is the scattered wave which is of interest here.

The layout for this subsection is:

1. We divide the scattering problem into 2 regimes:
   (a) Wavelength $\lesssim$ scatterer: This is called Diffraction and shall not be discussed in this book.
   (b) Wavelength $>$ scatterer: This will be called Scattering and it is the topic for this chapter.
2. We define the total scattering cross section σ.
3. We then show that there is a general relationship between the total cross section and forward scattering amplitude. This is called the optical theorem.
4. Finally, we work out 2 very basic models of scatterers.

5.1 Statement of the scattering problem

We consider a localised scattering region and an intensity flux of EM waves I_i incident on it.

Assume that there is a detector at (r, θ, ϕ) from the scatterer and it has an area of $dA = r^2 d\Omega$, so

$$\text{Power entering the detector} = I_s dA \tag{5.1}$$

but physically, this power should be proportional to I_i and the solid angle the detector subtends,

$$\implies I_s dA \propto I_i d\Omega \tag{5.2}$$
$$\implies I_s dA = A(\theta, \phi) I_i d\Omega \tag{5.3}$$
$$\mid \text{recall that } dA = r^2 d\Omega$$
$$\implies A(\theta, \phi) = r^2 \frac{I_s}{I_i} \tag{5.4}$$

where $A(\theta, \phi)$ is a "proportionality function". We see that $A(\theta, \phi)$ has the dimensions of area. It is called the differential cross section $A(\theta, \phi) = \dfrac{d\sigma}{d\Omega}$. The total cross section is thus defined as

$$\text{Total cross section:} \quad \sigma = \int A(\theta, \phi) d\Omega = \int \frac{d\sigma}{d\Omega} d\Omega \tag{5.5}$$

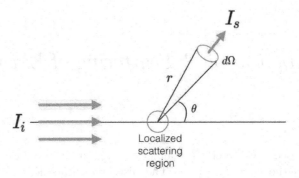

FIGURE 5.1
Setup for the definition of total cross section σ.

$$\sigma = \int_0^{2\pi} \int_0^{\pi} A(\theta,\phi) \sin\theta\, d\theta\, d\phi = \int_0^{2\pi} \int_0^{\pi} \frac{d\sigma}{d\Omega} \sin\theta\, d\theta\, d\phi \quad (5.6)$$

if $\frac{d\sigma}{d\Omega}$ is independent of ϕ

$$\sigma = 2\pi \int_0^{\pi} \frac{d\sigma}{d\Omega} \sin\theta\, d\theta \quad (5.7)$$

5.2 Formalism: Optical theorem

Again we set up a similar localised scattering geometry, as shown in Figure 5.2.

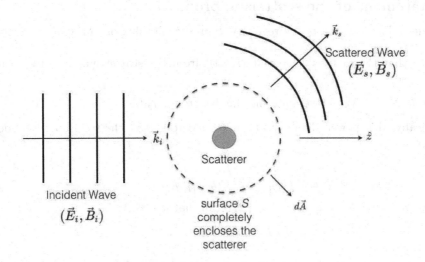

FIGURE 5.2
Scattering geometry for the derivation of optical theorem.

Standard Application 3: Scattering of EM waves

The total fields at all points in space are

$$\vec{E}_t = \vec{E}_i + \vec{E}_s \quad , \quad \vec{B}_t = \vec{B}_i + \vec{B}_s = \frac{1}{c}\hat{k}_i \times \vec{E}_i + \frac{1}{c}\hat{k}_s \times \vec{E}_s \tag{5.8}$$

We shall take this explicit (asymptotic) form for the total electric field.

$$\vec{E}_t = \underbrace{E_0 e^{i\vec{k}_i \cdot \vec{r}} e^{-i\omega t} \hat{x}}_{\vec{E}_i} + \underbrace{E_0 \frac{e^{ikr}}{r} \vec{F}(\theta, \phi) e^{-i\omega t}}_{\vec{E}_s} \tag{5.9}$$

This form assumes that $\vec{E}_i$ is an incoming plane wave, polarised in the x direction and we shall take the propagation direction to be $\vec{k}_i = k\hat{z}$. Then $\vec{E}_s$ is an outgoing scattered spherical wave with $\vec{k}_s = k\hat{r}$. The vector function $\vec{F}(\theta, \phi)$ is called the scattering amplitude function (which we shall see why later). We shall also assume that $\vec{F}$ is independent of ϕ, $\vec{F}(\theta, \phi) = \vec{F}(\theta)$ since we will only discuss spherically symmetric scatterers. We must note that since $\vec{E}_s \perp \vec{k}_s$, $\vec{k}_s \cdot \vec{F}(\theta) = 0$.

$$\vec{E}_t = \underbrace{E_0 e^{ikr\cos\theta} e^{-i\omega t} \hat{x}}_{\vec{E}_i} + \underbrace{E_0 \frac{e^{ikr}}{r} \vec{F}(\theta) e^{-i\omega t}}_{\vec{E}_s} \tag{5.10}$$

We shall now state the optical theorem now then we will carry out its proof.

$$\boxed{\text{Optical theorem:} \quad \sigma_t = \frac{4\pi}{k} \Im\left[\hat{x} \cdot \vec{F}(\theta = 0)\right]} \tag{5.11}$$

It states that the total cross section (which includes absorption cross section σ_a & scattering cross section σ_s) is proportional to the imaginary part (symbol $\Im$) of the forward scattering amplitude.

We start the proof by first proving a complex identity. We need this identity for 2 reasons: we are using the complex representation of EM waves and we are taking a time average for the power later.

$$\boxed{\text{Identity:} \quad \left\langle \Re(\vec{E}) \cdot \Re(\vec{B}) \right\rangle_{\text{time}} = \frac{1}{2}\Re\left(\vec{E} \cdot \vec{B}^*\right)} \tag{5.12}$$

where $\vec{E} = \vec{E}(\vec{r})e^{-i\omega t}$, $\vec{B} = \vec{B}(\vec{r})e^{-i\omega t}$ and the symbol $\Re$ means the "real part". Here is the proof for the identity,

$$\text{RHS} = \frac{1}{2}\Re\left(\vec{E} \cdot \vec{B}^*\right) = \frac{1}{2}\Re\left(\vec{E}(\vec{r}) \cdot \vec{B}^*(\vec{r})\right) \tag{5.13}$$

$$\frac{1}{2}\Re\left(\vec{E} \cdot \vec{B}^*\right) = \frac{1}{2}\Re\left[(\vec{E}_R + i\vec{E}_I) \cdot (\vec{B}_R - i\vec{B}_I)\right] \tag{5.14}$$

$$\frac{1}{2}\Re\left(\vec{E} \cdot \vec{B}^*\right) = \frac{1}{2}\left(\vec{E}_R \cdot \vec{B}_R + \vec{E}_I \cdot \vec{B}_I\right) \tag{5.15}$$

$$\text{LHS} = \left\langle \Re(\vec{E}) \cdot \Re(\vec{B}) \right\rangle_{\text{time}} \tag{5.16}$$

$$\text{LHS} = \frac{1}{T}\int_0^T \left\{ \frac{1}{2}\left(\vec{E}(\vec{r})e^{-i\omega t} + \vec{E}^*(\vec{r})e^{i\omega t}\right) \cdot \frac{1}{2}\left(\vec{B}(\vec{r})e^{-i\omega t} + \vec{B}^*(\vec{r})e^{i\omega t}\right) \right\} dt \tag{5.17}$$

$$\text{LHS} = \frac{1}{4}\vec{E}(\vec{r})\cdot\vec{B}^*(\vec{r}) + \frac{1}{4}\vec{E}^*(\vec{r})\cdot\vec{B}(\vec{r})$$

$$+ \frac{\omega}{2\pi}\frac{1}{4}\int_0^{2\pi/\omega}\left(\vec{E}(\vec{r})\cdot\vec{B}(\vec{r})e^{-2i\omega t} + \vec{E}^*(\vec{r})\cdot\vec{B}^*(\vec{r})e^{2i\omega t}\right)dt$$

$$\Big| \quad \text{then} \int_0^{2\pi/\omega} e^{\pm 2i\omega t}dt = \frac{1}{\pm 2i\omega}\left[e^{\pm 2i\omega t}\right]_0^{2\pi/\omega} = 0 \tag{5.18}$$

$$\text{LHS} = \frac{1}{4}\left[(\vec{E}_R + i\vec{E}_I)\cdot(\vec{B}_R - i\vec{B}_I) + (\vec{E}_R - i\vec{E}_I)\cdot(\vec{B}_R + i\vec{B}_I)\right] \tag{5.19}$$

$$\text{LHS} = \frac{1}{2}\left(\vec{E}_R\cdot\vec{B}_R + \vec{E}_I\cdot\vec{B}_I\right) \tag{5.20}$$

Hence the identity is proved and we will also use it in the case of cross product expressions.

We define the absorbed power as the inward flux (negative $d\vec{A}$) of the (total fields) Poynting vector over surface S.[1]

$$\langle P_a\rangle_{\text{time}} = -\frac{1}{\mu_0}\oint_S \left\langle \Re(\vec{E}_t)\times\Re(\vec{B}_t)\right\rangle_{\text{time}}\cdot d\vec{A} \tag{5.21}$$

$$\Big| \quad \text{use the complex identity, equation (5.12)}$$

$$\langle P_a\rangle_{\text{time}} = -\frac{1}{2\mu_0}\oint_S \Re(\vec{E}_t\times\vec{B}_t^*)\cdot d\vec{A} \tag{5.22}$$

The scattered power is defined as the outward flux of the (scattered fields) Poynting vector over surface S.

$$\langle P_s\rangle_{\text{time}} = \frac{1}{\mu_0}\oint_S \left\langle \Re(\vec{E}_s)\times\Re(\vec{B}_s)\right\rangle_{\text{time}}\cdot d\vec{A} \tag{5.23}$$

$$\Big| \quad \text{use the complex identity, equation (5.12)}$$

$$\langle P_s\rangle_{\text{time}} = \frac{1}{2\mu_0}\oint_S \Re(\vec{E}_s\times\vec{B}_s^*)\cdot d\vec{A} \tag{5.24}$$

The total power is thus

$$\langle P_t\rangle_{\text{time}} = \langle P_a\rangle_{\text{time}} + \langle P_s\rangle_{\text{time}} \tag{5.25}$$

$$\langle P_t\rangle_{\text{time}} = \frac{1}{2\mu_0}\oint_S \left[\Re(\vec{E}_s\times\vec{B}_s^*) - \Re(\vec{E}_t\times\vec{B}_t^*)\right]\cdot d\vec{A} \tag{5.26}$$

$$\langle P_t\rangle_{\text{time}} = -\frac{1}{2\mu_0}\oint_S \Re\left[\vec{E}_i\times\vec{B}_i^* + \vec{E}_i\times\vec{B}_s^* + \vec{E}_s\times\vec{B}_i^*\right]\cdot d\vec{A} \tag{5.27}$$

$$\Big| \quad \text{the first term: } \oint_S \Re\left[\vec{E}_i\times\left(\frac{1}{c}\hat{k}_i\times\vec{E}_i^*\right)\right]\cdot d\vec{A}$$

$$\Big| \qquad\qquad = \underbrace{\frac{1}{c}\oint_S |E_0|^2\hat{z}\cdot d\vec{A} = \frac{|E_0|^2}{c}\oint_S \hat{z}\cdot d\vec{A} = 0}_{\text{since } \hat{k}_i = \hat{z} \text{ and } \hat{x}\times(\hat{z}\times\hat{x}) = \hat{z}}$$

$$\langle P_t\rangle_{\text{time}} = -\frac{1}{2\mu_0}\oint_S \Re\left[\vec{E}_i\times\vec{B}_s^* + \vec{E}_s\times\vec{B}_i^*\right]\cdot d\vec{A} \tag{5.28}$$

$$\Big| \quad \text{they are interference terms between the incident and scattered waves}$$

$$\langle P_t\rangle_{\text{time}} = -\frac{1}{2\mu_0}\oint_S \Re\left[\vec{E}_i\times\left(\frac{1}{c}\hat{k}_s\times\vec{E}_s^*\right) + \vec{E}_s\times\left(\frac{1}{c}\hat{k}_i\times\vec{E}_i^*\right)\right]\cdot d\vec{A} \tag{5.29}$$

---

[1] By taking the total fields, we are including the interference between $\vec{E}_i$ and $\vec{E}_s$.

Standard Application 3: Scattering of EM waves

| use $\vec{E}_{i/s} \times (\hat{k}_{s/i} \times \vec{E}^*_{s/i}) = \underbrace{\hat{k}_{s/i}(\vec{E}_{i/s} \cdot \vec{E}_{s/i}) - \vec{E}^*_{s/i}(\vec{E}_{i/s} \cdot \hat{k}_{s/i})}_{\text{triple product identity}}$

| so, $\vec{E}_i \times (\hat{k}_s \times \vec{E}^*_s) = \dfrac{|E_0|^2}{r} e^{ikr\cos\theta - ikr}\left((\hat{x} \cdot \vec{F}^*)\hat{k}_s - (\hat{x} \cdot \hat{k}_s)\vec{F}^*\right)$

| and, $\vec{E}_s \times (\hat{k}_i \times \vec{E}^*_i) = \dfrac{|E_0|^2}{r} e^{-ikr\cos\theta + ikr}\left((\hat{x} \cdot \vec{F})\hat{z} - (\vec{F} \cdot \hat{z})\hat{x}\right)$

| We shall work with the first term first

| first term $= -\dfrac{1}{2\mu_0 c}\Re \oint_S \dfrac{|E_0|^2}{r} e^{ikr(\cos\theta - 1)}(\hat{x} \cdot \vec{F}^*(\theta))\hat{k}_s \cdot \hat{r} r^2 \sin\theta d\theta d\phi$

| but $\hat{k}_s = \hat{r}$ so first term $= -\dfrac{|E_0|^2}{2\mu_0 c}\Re \int_0^{2\pi} d\phi \int_0^{\pi} d\theta \sin\theta \, r e^{ikr(\cos\theta - 1)}(\hat{x} \cdot \vec{F}^*(\theta))$

| | for $\theta \neq 0$, $r \to$ large, $\underbrace{e^{ikr(\cos\theta - 1)}}_{\text{\& assuming } \vec{F}^*(\theta) \text{ is a smooth function}}$ is rapidly oscillating

| | and so any θ integration is zero even if integration range is small.

| | about the forward direction ($\theta \approx 0$), $\cos\theta - 1 \approx 0$, no oscillatory behaviour

| | thus the integral is not zero and to pick out that contribution,

| | we start by writing $\cos\theta \approx 1 - \dfrac{\theta^2}{2}$ and $\sin\theta \approx \theta$

| first term $\approx -\dfrac{|E_0|^2}{2\mu_0 c}\Re \int_0^{2\pi} d\phi (\hat{x} \cdot \vec{F}^*(\theta = 0) \int_0^{\delta} d\theta \, \theta r e^{ikr\left(-\frac{\theta^2}{2}\right)}$

| | where δ is a small upper limit and write $d\theta \, \theta = \dfrac{1}{2} d\theta^2$

| | first term $= -\dfrac{|E_0|^2}{2\mu_0 c}\Re \int_0^{2\pi} d\phi (\hat{x} \cdot \vec{F}^*(\theta = 0))\dfrac{1}{2}\left[r\left(-\dfrac{2}{ikr}\right) e^{-ikr\frac{\theta^2}{2}}\right]_0^{\delta}$

| | use mathematical prescription: $\lim\limits_{r \to \infty} e^{-ikr\frac{\delta^2}{2}} = $ average value $= 0$

| finally, first term $= -\dfrac{|E_0|^2}{2\mu_0 c}\dfrac{2\pi}{k}\Re\left(\dfrac{1}{i}(\hat{x} \cdot \vec{F}^*(\theta = 0)\right)$

| second term has $\hat{x} \cdot \hat{k}_s \xrightarrow{\text{forward limit}} \hat{x} \cdot \hat{z} = 0$

| third term has $\hat{z} \cdot \hat{r} \xrightarrow{\text{forward limit}} \hat{z} \cdot \hat{z} = 1$ and survives

| fourth term has $\hat{x} \cdot \hat{r} \xrightarrow{\text{forward limit}} \hat{x} \cdot \hat{z} = 0$

$$\langle P_t \rangle_{\text{time}} = -\dfrac{|E_0|^2}{2\mu_0 c}\dfrac{2\pi}{k}\Re\left(\dfrac{\hat{x} \cdot \vec{F}^*(\theta = 0)}{i} - \dfrac{\hat{x} \cdot \vec{F}(\theta = 0)}{i}\right) \tag{5.30}$$

$$\langle P_t \rangle_{\text{time}} = -\dfrac{|E_0|^2}{2\mu_0 c}\dfrac{2\pi}{k}\Re\left(-i\hat{x} \cdot (\vec{F}^*(\theta = 0) - \vec{F}(\theta = 0))\right) \tag{5.31}$$

| write $\vec{F}^*(\theta = 0) - \vec{F}(\theta = 0) = -2i\Im \vec{F}(\theta = 0)$

$$\langle P_t \rangle_{\text{time}} = \dfrac{|E_0|^2}{2\mu_0 c}\dfrac{4\pi}{k}\Im\left(\hat{x} \cdot \vec{F}(\theta = 0)\right) \tag{5.32}$$

To[2] see how $\langle P_t \rangle_{\text{time}}$ is related to σ_t, we work with

$$\langle P_s \rangle_{\text{time}} = \frac{1}{2\mu_0} \oint_S \Re\left(\vec{E}_s \times \vec{B}_s^*\right) \cdot d\vec{A} \tag{5.33}$$

$$\langle P_s \rangle_{\text{time}} = \frac{1}{2\mu_0} \oint_S \Re\left(\vec{E}_s \times \left(\frac{1}{c}\hat{k}_s \times \vec{E}_s^*\right)\right) \cdot d\vec{A} \tag{5.34}$$

$$\langle P_s \rangle_{\text{time}} = \frac{|E_0|^2}{2\mu_0 c} \oint_S \Re\left(\frac{1}{r^2}\left(\vec{F} \times (\hat{k}_s \times \vec{F}^*)\right)\right) \cdot d\vec{A} \tag{5.35}$$

| write $\vec{F} \times (\hat{k}_s \times \vec{F}^*) = \hat{k}_s(\vec{F} \cdot \vec{F}^*) - \vec{F}^* \underbrace{(\vec{F} \cdot \hat{k}_s)}_{=0} = |\vec{F}|^2 \hat{k}_s$,

| and $\hat{k}_s = \hat{r}$ and $d\vec{A} = \hat{r} r^2 \sin\theta d\theta d\phi$

$$\langle P_s \rangle_{\text{time}} = \frac{|E_0|^2}{2\mu_0 c} \int_0^{2\pi} d\phi \int_0^{\pi} d\theta \sin\theta |\vec{F}|^2 \tag{5.36}$$

| recall $\dfrac{d\sigma_s}{d\Omega} = r^2 \dfrac{I_s}{I_i} = r^2 \dfrac{\frac{|E_0|^2}{r^2}|\vec{F}|^2}{|E_0|^2} = |\vec{F}|^2$

| The statement $\dfrac{d\sigma_s}{d\Omega} = |\vec{F}|^2$ is why we call $\vec{F}$ the scattering amplitude.

$$\langle P_s \rangle_{\text{time}} = \frac{|E_0|^2}{2\mu_0 c} \int_0^{2\pi} d\phi \int_0^{\pi} d\theta \sin\theta \frac{d\sigma_s}{d\Omega} \tag{5.37}$$

$$\langle P_s \rangle_{\text{time}} = \frac{|E_0|^2}{2\mu_0 c} \sigma_s \tag{5.38}$$

Thus we can say that

$$\langle P_t \rangle_{\text{time}} = \frac{|E_0|^2}{2\mu_0 c} \frac{4\pi}{k} \Im\left(\hat{x} \cdot \vec{F}(\theta = 0)\right) \tag{5.39}$$

$$\implies \frac{|E_0|^2}{2\mu_0 c} \sigma_t = \frac{|E_0|^2}{2\mu_0 c} \frac{4\pi}{k} \Im\left(\hat{x} \cdot \vec{F}(\theta = 0)\right) \tag{5.40}$$

$$\implies \sigma_t = \frac{4\pi}{k} \Im\left(\hat{x} \cdot \vec{F}(\theta = 0)\right) \tag{5.41}$$

which is the optical theorem stated earlier. We should think for a while why would there be a relationship between the total cross section and the forward scattering amplitude. In forward scattering, the scattered spherical wave and the incident plane wave will align and interfere. For a "rather solid" (large σ_t) scatterer, the forward flux will be smaller due to the interference. Thus, by the interference argument, we expect a relationship between σ_t and $\vec{F}(\theta = 0)$.[3]

5.3 Basic scattering models

5.3.1 Scattering by free charge: Thomson scattering

The model of a scatterer is a single free charged particle of charge q and mass m. The incoming plane wave propagates in the x direction and it is polarised in the z direction.

$$\vec{E}_i = E_0 e^{i\vec{k}\cdot\vec{r}} e^{-i\omega t} \hat{z} \quad \text{where } \vec{k} = k\hat{x} \text{ and } \vec{k} \cdot \vec{r} = kr\sin\theta\cos\phi \tag{5.42}$$

---

[2] The following working is based on "Principles of Quantum Mechanics 2nd Edition" page 555 by R. Shankar.
[3] This argument is taken from [1] page 114.

Standard Application 3: Scattering of EM waves

The (non-relativistic) equation of motion of the charged particle is

$$m\frac{d\vec{v}}{dt} = q\Re(\vec{E}_i) \quad \text{(we assume that } v \ll c) \tag{5.43}$$

so the charged particle oscillates up and down the z axis about the origin. The (time-averaged) scattered power is given by Larmor's formula (for a slow moving but accelerating point charge).

$$\frac{dP_s}{d\Omega} = \frac{\mu_0 q^2}{16\pi^2 c}\langle a^2\rangle_{\text{time}} \sin^2\theta \tag{5.44}$$

$$\frac{dP_s}{d\Omega} = \frac{\mu_0 q^2}{16\pi^2 c}\left(\frac{q}{m}\right)^2 \left\langle \Re(\vec{E}_i)\cdot\Re(\vec{E}_i)\right\rangle_{\text{time}} \sin^2\theta \tag{5.45}$$

| use complex identity, eq(5.12): $\left\langle \Re(\vec{E}_i)\cdot\Re(\vec{E}_i)\right\rangle_{\text{time}} = \frac{1}{2}\vec{E}_i\cdot\vec{E}_i^* = \frac{1}{2}|E_0|^2$

$$\frac{dP_s}{d\Omega} = \frac{\mu_0 q^4}{16\pi^2 c m^2}\frac{|E_0|^2}{2}\sin^2\theta \tag{5.46}$$

Total Power: $P_s = \dfrac{\mu_0 q^2}{6\pi c}a^2 \tag{5.47}$

| from eqs (5.44) and (5.46), we can see $\langle a^2\rangle_{\text{time}} = \dfrac{q^2}{m^2}\dfrac{|E_0|^2}{2}$

$$P_s = \frac{\mu_0 q^4}{6\pi c m^2}\frac{|E_0|^2}{2} \tag{5.48}$$

To calculate the scattering cross section $\dfrac{d\sigma}{d\Omega} = r^2 \dfrac{I_s}{I_i}$, we need the scattered intensity

$$I_s = \frac{dP_s}{dA} = \frac{\frac{dP_s}{d\Omega}}{\frac{dA}{d\Omega}} = \frac{\frac{dP_s}{d\Omega}}{r^2} = \frac{1}{r^2}\frac{dP_s}{d\Omega} \tag{5.49}$$

and the incident intensity,

$$I_i = \left|\langle\vec{S}_i\rangle_{\text{time}}\right| \tag{5.50}$$

$$I_i = \frac{1}{\mu_0}\left|\langle\Re(\vec{E}_i)\times\Re(\vec{B}_i)\rangle_{\text{time}}\right| \tag{5.51}$$

| use the complex identity $\langle\Re(\vec{E}_i)\times\Re(\vec{B}_i)\rangle_{\text{time}} = \frac{1}{2}\vec{E}_i\times\vec{B}_i^* = \frac{1}{2}\left(\vec{E}_i\times\left(\frac{1}{c}\hat{k}_i\times\vec{E}_i^*\right)\right)$

$$I_i = \frac{1}{2\mu_0 c}\left|\left(\vec{E}_i\times\left(\frac{1}{c}\hat{k}_i\times\vec{E}_i^*\right)\right)\right| \tag{5.52}$$

| use vector identity $\vec{E}_i\times\left(\hat{k}_i\times\vec{E}_i^*\right) = \hat{k}_i|\vec{E}_i|^2 - \vec{E}_i^*(\hat{k}_i\cdot\vec{E}_i)$ and $\hat{k}_i\cdot\vec{E}_i = 0$

$$I_i = \frac{1}{2\mu_0 c}\left|\hat{k}_i|\vec{E}_i|^2\right|$$

$$\boxed{I_i = \frac{1}{2\mu_0 c}|E_0|^2} \tag{5.53}$$

Differential cross section: $\dfrac{d\sigma}{d\Omega} = r^2\dfrac{I_s}{I_i} = r^2\dfrac{\frac{1}{r^2}\frac{dP_s}{d\Omega}}{\frac{1}{2\mu_0 c}|E_0|^2} = \dfrac{\mu_0^2 q^4}{16\pi^2 m^2}\sin^2\theta \tag{5.54}$

Total cross section (Thomson cross section):
$$\sigma_{Th} = 2\pi \int_0^\pi \frac{d\sigma}{d\Omega} \sin\theta d\theta \tag{5.55}$$

$$\sigma_{Th} = \frac{\mu_0^2 q^4}{8\pi m^2} \int_0^\pi \sin^3\theta d\theta \tag{5.56}$$

$$\left|\text{ recall that } \int_0^\pi \sin^3\theta d\theta = \frac{4}{3}\right.$$

$$\boxed{\sigma_{Th} = \frac{\mu_0^2 q^4}{6\pi m^2}} \tag{5.57}$$

This[4] scattering cross section is independent of frequency. However, it is only valid for low frequencies where the momentum of the photon (which is $\frac{E}{c}$) is small. When the momentum of the photon is comparable to mc (of the electron), this is Compton scattering and the scattering cross section is given by the Klein–Nishina formula (derived from the quantum field theory of Electrodynamics (QED)).

5.3.2 Scattering by bound charge

The model of a scatterer is a charge (usually the electron) being bounded by a damped harmonic oscillator force.[5] The equation of motion is now

$$m\frac{d^2\vec{r}}{dt^2} + m\omega_0^2\vec{r} + m\gamma\frac{d\vec{r}}{dt} = q\Re(\vec{E}_i) \quad \text{(we assume } v \ll c\text{)} \tag{5.58}$$

and ω_0 is the natural frequency of the harmonic force, γ is the damping constant[6] and $\vec{E}_i$ is the (complex) incident plane wave $\vec{E}_i = E_0 e^{i\vec{k}\cdot\vec{r}} e^{-i\omega t} \hat{z}$. We solve it as follows:

Imagine another problem: $m\frac{d^2\vec{y}}{dt^2} + m\omega_0^2\vec{y} + m\gamma\frac{d\vec{y}}{dt} = q\Im(\vec{E}_i)$ (5.59)

multiply i and add the 2 equations together |

$$\left(m\frac{d^2}{dt^2} + m\omega_0^2 + m\gamma\frac{d}{dt}\right)(\vec{r} + i\vec{y}) = q(\Re(\vec{E}_i) + i\Im(\vec{E}_i)) \tag{5.60}$$

call $\vec{r} + i\vec{y} = \vec{z}$ |

$$m\frac{d^2\vec{z}}{dt^2} + m\omega_0^2\vec{z} + m\gamma\frac{d\vec{z}}{dt} = q\vec{E}_i \tag{5.61}$$

insert solution ansatz $\vec{z} = \vec{z}_0 e^{-i\omega t}$ |

$$-m\omega^2\vec{z}_0 e^{-i\omega t} + m\omega_0^2\vec{z}_0 e^{-i\omega t} - im\omega\gamma\vec{z}_0 e^{-i\omega t} = q\vec{E}_i \tag{5.62}$$

$$\vec{r}(t) = \Re(\vec{z}) = \Re(\vec{z}_0 e^{-i\omega t}) = \frac{q}{m}\Re\left(\frac{\vec{E}_i}{\omega_0^2 - \omega^2 - i\gamma\omega}\right) \tag{5.63}$$

Just like in the previous section, we need the time-averaged acceleration.

$$\langle a^2 \rangle_{\text{time}} = \left\langle \frac{d^2\vec{r}}{dt^2} \cdot \frac{d^2\vec{r}}{dt^2} \right\rangle_{\text{time}} \tag{5.64}$$

---

[4] For historical reasons, σ_{Th} is always written as $\sigma_{Th} = \frac{\mu_0^2 q^4}{6\pi m^2} = \frac{8\pi}{3}r_e^2$ where $r_e^2 = \frac{q^2}{4\pi\epsilon_0 mc^2}$ and r_e is known as the classical radius of the electron.

[5] This harmonic force could be the harmonic approximation about the minimum of the potential energy of the electron-nucleus system. The damping could be due to the energy dissipation of an accelerating charged particle that radiates.

[6] Do not confuse it with the Lorentz γ factor.

Standard Application 3: Scattering of EM waves

$$\langle a^2 \rangle_{\text{time}} = \frac{q^2}{m^2} \left\langle \Re\left(\frac{\ddot{\vec{E}}_i}{\omega_0^2 - \omega^2 - i\gamma\omega}\right) \cdot \Re\left(\frac{\ddot{\vec{E}}_i}{\omega_0^2 - \omega^2 - i\gamma\omega}\right) \right\rangle_{\text{time}} \quad (5.65)$$

$\qquad$ | use the complex identity, eq (5.12) $\langle \Re(\vec{A}) \cdot \Re(\vec{B}) \rangle_{\text{time}} = \frac{1}{2} \vec{A} \cdot \vec{B}^*$

$$\langle a^2 \rangle_{\text{time}} = \frac{q^2}{m^2} \frac{1}{2} \frac{\ddot{\vec{E}}_i \cdot \ddot{\vec{E}}_i^*}{(\omega_0^2 - \omega^2 - i\gamma\omega)(\omega_0^2 - \omega^2 + i\gamma\omega)} \quad (5.66)$$

$\qquad$ | use $\vec{E}_i = E_0 e^{i\vec{k}\cdot\vec{r}} e^{-i\omega t} \hat{z}$ so $\ddot{\vec{E}}_i = -\omega^2 \vec{E}_i$ and $\ddot{\vec{E}}_i \cdot \ddot{\vec{E}}_i^* = \omega^4 |E_0|^2$

$$\langle a^2 \rangle_{\text{time}} = \frac{q^2}{m^2} \frac{|E_0|^2}{2} \frac{\omega^4}{(\omega_0^2 - \omega^2)^2 + (\gamma\omega)^2} \quad (5.67)$$

Scattered power per solid angle,

$$\frac{dP_s}{d\Omega} = \frac{\mu_0 q^2}{16\pi^2 c} \langle a^2 \rangle_{\text{time}} \sin^2\theta = \frac{\mu_0 q^4}{16\pi^2 c m^2} \frac{|E_0|^2}{2} \frac{\omega^4}{(\omega_0^2 - \omega^2)^2 + (\gamma\omega)^2} \sin^2\theta \quad (5.68)$$

Total scattered power,

$$P_s = \frac{\mu_0 q^2}{6\pi c} \langle a^2 \rangle_{\text{time}} = \frac{\mu_0 q^4}{6\pi c m^2} \frac{|E_0|^2}{2} \frac{\omega^4}{(\omega_0^2 - \omega^2)^2 + (\gamma\omega)^2} \quad (5.69)$$

Recall equation (5.49): $I_s = \frac{1}{r^2} \frac{dP_s}{d\Omega}$ and equation (5.53): $I_i = \frac{1}{2\mu_0 c} |E_0|^2$ and so the scattering cross section $\frac{d\sigma}{d\Omega}$ is

$$\frac{d\sigma}{d\Omega} = r^2 \frac{I_s}{I_i} = r^2 \frac{\frac{1}{r^2}\frac{dP_s}{d\Omega}}{\frac{1}{2\mu_0 c}|E_0|^2} = \frac{\mu_0^2 q^4}{16\pi^2 m^2} \frac{\omega^4}{(\omega_0^2 - \omega^2)^2 + (\gamma\omega)^2} \sin^2\theta \quad (5.70)$$

Total cross section,

$$\sigma = 2\pi \int_0^\pi \frac{d\sigma}{d\Omega} d\Omega = \frac{\mu_0^2 q^4}{8\pi m^2} \frac{\omega^4}{(\omega_0^2 - \omega^2)^2 + (\gamma\omega)^2} \int_0^\pi \sin^3\theta \, d\theta \quad (5.71)$$

$\qquad$ | with $\int_0^\pi \sin^3\theta \, d\theta = \frac{4}{3}$ and $\sigma_{\text{Th}} = \frac{\mu_0^2 q^4}{6\pi m^2}$

$$\boxed{\sigma = \sigma_{\text{Th}} \frac{\omega^4}{(\omega_0^2 - \omega^2)^2 + (\gamma\omega)^2}} \quad (5.72)$$

We shall look at 3 regions of interest:

Region 1: $\omega \gg \omega_0, \gamma$ The incident EM wave has a very small period and the internal harmonic force and dissipation do not set in that quickly, so the electron behaves as if it were free and the scattering cross section should approach σ_{Th}.

$$\sigma = \sigma_{\text{Th}} \frac{\omega^4}{(\omega_0^2 - \omega^2)^2 + (\gamma\omega)^2} \longrightarrow \sigma_{\text{Th}} \frac{\omega^4}{(-\omega^2)^2} = \sigma_{\text{Th}} \quad \text{(as expected)} \quad (5.73)$$

Region 2: $\omega \ll \omega_0$

$$\sigma = \sigma_{\text{Th}} \frac{\omega^4}{(\omega_0^2 - \omega^2)^2 + (\gamma\omega)^2} \longrightarrow \sigma_{\text{Th}} \frac{\omega^4}{\omega_0^4} = \sigma_{\text{Rayleigh}} \quad (5.74)$$

This is called Rayleigh cross section. Thus the highest frequencies are scattered the most as they "see" a larger scattering cross section.[7]

Region 3: resonant scattering $\omega = \omega_0$

$$\sigma = \sigma_{\text{Th}} \frac{\omega_0^4}{(\gamma \omega_0)^2} = \sigma_{\text{Th}} \frac{\omega_0^2}{\gamma^2} \qquad (5.75)$$

For small damping, i.e. $\omega_0 \gg \gamma$, the cross section is much larger than σ_{Th}.

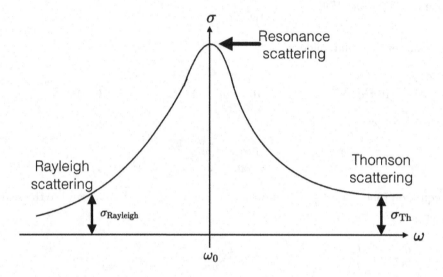

FIGURE 5.3
Graphical sketch of the ω dependence of the total cross section σ.

Closing Remarks:

- The next example, "scattering by dielectric sphere", requires much more work to derive, in particular, the mathematics of "vector spherical harmonics" is needed.

- We have essentially covered the theoretical groundwork behind Electrical and Electronic Engineering (EEE) engineering and you can proceed to read about applications in EEE engineering.

- For progress in theoretical physics, we need to change the mathematical language from vector calculus to tensor calculus in Special Relativity. This is the aim of Part II.

---

[7] Finally now you have a basic understanding of why the sky is blue.

6

Exercises for Part I

6.1 Maxwell stress tensor

Find the net force that the southern hemisphere of a uniformly charged sphere exerts on the northern hemisphere. Use the Maxwell stress tensor to calculate. Express your answer in terms of the radius R and the total charge Q. (Hint: See page of [4]) (Answer: $F = \frac{1}{4\pi\epsilon_0} \frac{3Q^2}{16R^2}$)

6.2 Existence of the gauges

The gauge choices exist if the corresponding gauge functions λ exist.

1. **Coulomb gauge:** For Coulomb gauge, work out the (differential) equation that the gauge function λ satisfies. It should be a familiar equation to you, so write down the solution for λ directly.

2. **Lorenz gauge:** For the Lorenz gauge, work out the (differential) equation that the gauge function λ satisfies. It turns out that the gauge function λ is not unique. If we have $\bar{\lambda} = \lambda + f$, what differential equation does function f satisfy so that λ and $\bar{\lambda}$ satisfy the answer you wrote earlier? Comment on the form of function f.

6.3 Check back the retarded potential solutions

Substitute the retarded potential

$$\phi(\vec{r}, t) = \frac{1}{4\pi\epsilon_0} \int \frac{\rho(\vec{r}', t_r)}{|\vec{r} - \vec{r}'|} dV' \tag{6.1}$$

into

$$\left(\vec{\nabla}^2 - \frac{1}{c^2} \frac{\partial^2}{\partial t^2}\right) \phi = -\frac{\rho(\vec{r}, t)}{\epsilon_0} \tag{6.2}$$

and check it is satisfied.

6.4 Retarded potentials satisfy Lorenz gauge

Check that the retarded potentials, equation (6.1) and

$$\vec{A}(\vec{r},t) = \frac{\mu_0}{4\pi} \int \frac{\vec{J}(\vec{r}',t_r)}{|\vec{r}-\vec{r}'|} dV' \qquad (6.3)$$

satisfies the Lorenz gauge

$$\vec{\nabla} \cdot \vec{A} = -\mu_0 \epsilon_0 \frac{\partial \phi}{\partial t} \qquad (6.4)$$

6.5 Quasistatic approximation of Jefimenko's equation

Suppose the current density $\vec{J}$ changes slowly enough so that we can ignore all higher derivative terms in the Taylor expansion

$$\vec{J}(\vec{r}',t_r) = \vec{J}(\vec{r}',t) + (t_r - t)\dot{\vec{J}}(\vec{r}',t) + \cdots \qquad (6.5)$$

Show that a nice cancellation occurs in the Jefimenko equation

$$\vec{B}(\vec{r},t) = \frac{\mu_0}{4\pi} \int \left(\frac{\vec{J}(\vec{r}',t_r)}{R^2} + \frac{\dot{\vec{J}}(\vec{r}',t_r)}{cR} \right) \times \hat{R} dV' \qquad (6.6)$$

to give

$$\vec{B}(\vec{r},t) = \frac{\mu_0}{4\pi} \int \frac{\vec{J}(\vec{r}',t) \times \hat{R}}{R^2} dV'$$

which is Biot-Savart's law at the present time. It means that in first order, the effect of retarded time and the effect of time-dependent current density cancel each other.

6.6 Charge moving at constant velocity

For the E-field of a charge moving at constant velocity,

$$\vec{E} = \frac{q}{4\pi\epsilon_0} \frac{1 - \frac{v^2}{c^2}}{\left(1 - \frac{v^2}{c^2}\sin^2\alpha\right)^{3/2}} \frac{\hat{r}_p}{r_p^2} \qquad (6.7)$$

1. Write down the expression of $E^\|$ which is the component of $\vec{E}$ parallel to the trajectory. Sketch 2 curves on the $E^\|$ vs vt axes, (i) for $v \ll c$ and (ii) for $v \approx c$. On the graph, the relationship between the maxima/minima of the 2 curves must be sketched accurately.

2. Write down the expression of $E^\perp$ which is the component of $\vec{E}$ perpendicular to the trajectory. Then show that $\frac{E^\|}{E^\perp} = \frac{x - vt}{y}$. Sketch 2 curves on the $E^\perp$ vs vt axes, (i) for $v \ll c$ and (ii) for $v \approx c$.

6.7 Decay of the classical atom

This question is historically an important calculation in the development of quantum mechanics. Assume that the hydrogen atom in ground state is the electron in a circular orbit of 5×10^{-11} m around the proton. Assume that only Coulomb attraction is involved. According to electrodynamics, the electron radiates because it is accelerating and hence it will lose energy and spiral into the nucleus. Show that $v \ll c$ at the beginning and use Larmor's formula to calculate the time it takes for the electron to reach $r = 0$. (Answer: 1.3×10^{-11}s)

6.8 Dipole retarded potentials satisfy Lorenz gauge

For the Hertzian dipole, the retarded potentials are

$$\phi(r,\theta,t) = \frac{p_0 \cos\theta}{4\pi\epsilon_0 r} \left\{ -\frac{\omega}{c} \sin\left(\omega\left(t - \frac{r}{c}\right)\right) + \frac{1}{r} \cos\left(\omega\left(t - \frac{r}{c}\right)\right) \right\} \qquad (6.8)$$

$$\vec{A}(r,\theta,t) = -\frac{\mu_0 p_0 \omega}{4\pi r} \sin\left(\omega\left(t - \frac{r}{c}\right)\right) \hat{z} \qquad (6.9)$$

check that they satisfy the Lorenz gauge. Do not use the far field approximation.

6.9 Recover various cases

Using the expression of total power for an arbitrary distribution

$$P = \frac{\mu_0 \ddot{p}(\hat{t})^2}{6\pi c} \qquad (6.10)$$

recover the previous results of total power for

1. Hertzian dipole.
2. (Slow moving) point charge.

6.10 Magnetic dipole radiation

Suppose we have a current loop of radius b with alternating current $I(t) = I_0 \cos\omega t$ and its centre is set at the origin with its plane in the xy plane, calculate the radiative $\vec{E}$ and $\vec{B}$ fields. State all approximations clearly. (Hint: see page 451 of [4]) Then,

1. Calculate the time-averaged total power.
2. Show that the ratio of the power of this magnetic dipole to the power of the Hertzian dipole is $\left(\frac{\omega b}{c}\right)^2$ by setting the electric dipole length d to be $d = \pi b$.
3. Give a reason why this setup has no electric dipole radiation.

6.11 Thomson scattering is nonrelativistic

The scattering cross section for Compton scattering is the Klein-Nishina formula

$$\frac{d\sigma}{d\Omega} = \alpha^2 \frac{\hbar^2}{4m^2c^2} \frac{\omega'^2}{\omega^2} \left[\frac{\omega'}{\omega} + \frac{\omega}{\omega'} + 4\sin^2\theta - 2 \right] \quad (6.11)$$

where $\alpha = \frac{\mu_0 c q^2}{4\pi\hbar}$ is the so-called fine structure constant and the Compton shift is $\omega' = \frac{\omega}{1 + \frac{\hbar\omega}{mc^2}(1-\cos\theta)}$. In the nonrelativistic limit, $\omega \to 0$ and $\omega' = \omega$, show that Thomson scattering cross section is obtained. (Notice that $\hbar$ terms are non-existent in Thomson cross section which means Thomson cross section does not take into account quantum effects.)

6.12 Satisfies optical theorem

The cross section calculated in the lecture notes for the bound scatterer

$$\sigma = \sigma_{\text{Th}} \frac{\omega^4}{(\omega_0^2 - \omega^2)^2 + (\gamma\omega)^2} \quad (6.12)$$

is just the scattered cross section. Let's obtain the total cross section σ_{total} by 2 ways:

1. Using the (far field) electric field of an arbitrary distribution (in Part I), $\vec{E}(r,\theta,t) = \frac{\mu_0 \ddot{p}(\tilde{t})}{4\pi} \frac{\sin\theta}{r} \hat{\theta}$, work out the scattering amplitude function $\vec{F}(\theta)$ and hence work out σ_{total} using the optical theorem.

2. Then work out the same σ_{total} by first working out $\langle P_{\text{total}} \rangle_{\text{time}}$ which is the time-averaged power that the dissipative force is drawing from the incident wave.

You should obtain $\sigma_{\text{total}} = \frac{\mu_0 q^2 c}{m} \frac{\gamma\omega^2}{(\omega_0^2 - \omega^2)^2 + (\gamma\omega)^2}$ in both cases.

6.13 Dielectric sphere scatterer

A dielectric sphere of radius a, in a static electric field acquires a dipole moment

$$\vec{p} = 4\pi\epsilon_0 \frac{\epsilon_2 - \epsilon_1}{\epsilon_2 + 2\epsilon_1} a^3 \vec{E} \quad (6.13)$$

where ϵ_2 is the dielectric constant of the dielectric sphere and ϵ_1 is the dielectric constant of the medium outside the dielectric sphere. Suppose a slowly varying (i.e. $\lambda \gg a$) plane EM wave is incident on the dielectric sphere, show that the total scattering cross section is

$$\sigma = \frac{8\pi}{3} k^4 a^6 \left(\frac{\epsilon_2 - \epsilon_1}{\epsilon_2 + 2\epsilon_1} \right)^2 \quad (6.14)$$

where $k = \frac{2\pi}{\lambda}$.

Part II

Electrodynamics in Lorentz Covariant Form (in tensor calculus form)

7

Special Relativity

7.1 Derivation of Lorentz transformation

Einstein's axioms for special relativity are:

1. Laws of nature and the results of all experiments are equivalent in all inertial frames. Inertial frames are reference frames at relative constant velocities to one another.[1]
2. Speed of light has the same value in all inertial frames and is independent of the motion of the light source. Speed of light is also the upper limit of physical entities.[2]

Note that from axiom 1, we can say that space is isotropic (all spatial directions are equivalent) and spacetime is homogeneous (origin of spacetime can be chosen arbitrarily). This will mean that the transformation is linear.

In fact, without axiom 2, we have the familiar Galilean transformation in Newtonian mechanics. It is a linear transformation.

Assuming the origins coincide at $t = 0$, we have

$$x = x' + ut \quad \xrightarrow{\text{inverse}} \quad x' = x - ut \tag{7.1}$$

$$\implies \frac{dx}{dt} = \frac{dx'}{dt} + u \tag{7.2}$$

$$\implies v = v' + u \tag{7.3}$$

which is the usual relative velocity expression.

Now we start with a more general linear transformation and incorporate axiom 2 towards the end.[3]

$$x' = \gamma(u)(x - ut) \tag{7.4}$$
$$y' = \alpha(u)y \tag{7.5}$$
$$z' = \alpha(u)z \tag{7.6}$$
$$t' = \mu(u)t + \epsilon(u)x \tag{7.7}$$

The coefficients of y' and z' are the same due to isotropy. Again due to isotropy, flipping x, x' and u will result in no change (we flipped y and y' so that we maintain a right handed

---

[1] This axiom is already in Newton's First Law.
[2] This axiom is new.
[3] "Linear" means power 1.

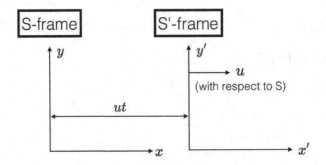

FIGURE 7.1
Two inertial frames with S'-frame at speed u relative to S-frame.

coordinate system).

$$-x' = \gamma(-u)(-x+ut) \tag{7.8}$$
$$-y' = -\alpha(-u)y \tag{7.9}$$
$$z' = \alpha(-u)z \tag{7.10}$$
$$t' = \mu(-u)t - \epsilon(-u)x \tag{7.11}$$

We compare to get $\gamma(-u) = \gamma(u)$, $\alpha(-u) = \alpha(u)$, $\mu(-u) = \mu(u)$ and $\epsilon(-u) = -\epsilon(u)$ so only ϵ is an odd function. We write $\epsilon(u)$ into this form where $\eta(u)$ is even.

$$\epsilon(u) = -\frac{u}{\eta(u)}\mu(u) \tag{7.12}$$

so the set of transformation equations are

$$x' = \gamma(u)(x-ut) \tag{7.13}$$
$$y' = \alpha(u)y \tag{7.14}$$
$$z' = \alpha(u)z \tag{7.15}$$
$$t' = \mu(u)\left(t - \frac{u}{\eta(u)}x\right) \tag{7.16}$$

We can "jump" into S'-frame and by axiom 1, the equations must have the same form. We interchange primed and unprimed coordinates and reverse the sign of u.

$$x = \gamma(u)(x'+ut') \tag{7.17}$$
$$y = \alpha(u)y' \tag{7.18}$$
$$z = \alpha(u)z' \tag{7.19}$$
$$t = \mu(u)\left(t' + \frac{u}{\eta(u)}x'\right) \tag{7.20}$$

Substitute $y' = \alpha(u)y$ to get $y = \alpha(u)^2 y$ so we choose $\alpha = 1$ \quad (7.21)

Then we proceed further with the x and t equations,

$$x' = \gamma(x-ut) \xrightarrow{\times \mu} x'\mu = \mu\gamma x - \mu\gamma ut \tag{7.22}$$
$$t' = \mu\left(t - \frac{u}{\eta}x\right) \xrightarrow{\times u\gamma} ut'\gamma = u\gamma\mu t - \frac{u^2}{\eta}\gamma\mu x \tag{7.23}$$

Special Relativity

$$\text{Eliminate } t \implies x'\mu + ut'\gamma = \mu\gamma x - \frac{u^2}{\eta}\gamma\mu x \tag{7.24}$$

$$x = \frac{x'\mu + u\gamma t'}{\mu\gamma - \frac{u^2}{\eta}\mu\gamma} \tag{7.25}$$

$$x = \frac{x'}{\gamma\left(1 - \frac{u^2}{\eta}\right)} + \frac{ut'}{\mu\left(1 - \frac{u^2}{\eta}\right)} \tag{7.26}$$

Compare with $x = \gamma(x' + ut') = \gamma x' + \gamma u t'$ to get $\gamma = \dfrac{1}{\gamma\left(1 - \frac{u^2}{\eta}\right)}$ and $\gamma = \dfrac{1}{\mu\left(1 - \frac{u^2}{\eta}\right)}$ so that $\gamma = \mu$ and $\gamma = \dfrac{1}{\sqrt{1 - \frac{u^2}{\eta(u)}}}$. The positive root is chosen so that when $u \to 0$, $\gamma \to 1$ and $x' \to x$.

The transformation equations are now,[4]

$$x' = \gamma(x - ut) \tag{7.27}$$
$$y' = y \tag{7.28}$$
$$z' = z \tag{7.29}$$
$$t' = \gamma\left(t - \frac{u}{\eta}x\right) \tag{7.30}$$

We determine η by using axiom 2 where every inertial frame sees speed of light as c. So we need the velocity addition formula. Take infinitesimal intervals,

$$dx' = \gamma(dx - u\,dt) \tag{7.31}$$
$$dt' = \gamma\left(dt - \frac{u}{\eta}dx\right) \tag{7.32}$$

divide one equation by the other,
$$\frac{dx'}{dt'} = \frac{dx - u\,dt}{dt - \frac{u}{\eta}dx} \tag{7.33}$$

$$\frac{dx'}{dt'} = \frac{\frac{dx}{dt} - u}{1 - \frac{u}{\eta}\frac{dx}{dt}} \tag{7.34}$$

$$v' = \frac{v - u}{1 - \frac{uv}{\eta}} \tag{7.35}$$

where v' is the velocity of the object measured in S'-frame and v is the velocity of the object measured in S-frame. Now consider the object to be light, then $v = c$ and $v' = c$, so

$$c = \frac{c - u}{1 - \frac{uc}{\eta}} \tag{7.36}$$

$$\eta = c^2 \tag{7.37}$$

---

[4] Note that if $\eta = \infty$ we get back the Galilean transformations. Thus, in simple terms, the difference between Newtonian mechanics and Special Relativity is that, the upper speed limit for Special Relativity is c and the upper speed limit for Newtonian Mechanics is ∞.

Finally with $\boxed{\gamma = \dfrac{1}{\sqrt{1 - \dfrac{u^2}{c^2}}}}$, the Lorentz transformation equations are,

$$\boxed{\begin{aligned} x' &= \gamma(x - ut) \\ y' &= y \\ z' &= z \\ t' &= \gamma\left(t - \frac{u}{c^2}x\right) \end{aligned}}$$
(7.38)
(7.39)
(7.40)
(7.41)

We generalise this for arbitrary relative velocities. We denote $\vec{u}$ to be the relative velocity in an arbitrary direction. We resolve the position vector into a component parallel to $\vec{u}$ and a component perpendicular to $\vec{u}$.

$$\vec{x} \cdot \vec{u} = ux\cos\theta = ux_\parallel \tag{7.42}$$

$$x_\parallel = \frac{\vec{x} \cdot \vec{u}}{u} \tag{7.43}$$

$$\vec{x}_\parallel = \frac{\vec{x} \cdot \vec{u}}{u}\hat{u} \tag{7.44}$$

$$= \frac{\vec{x} \cdot \vec{u}}{u^2}\vec{u} \tag{7.45}$$

$$\text{hence, } \vec{x}_\perp = \vec{x} - \vec{x}_\parallel = \vec{x} - \frac{\vec{x} \cdot \vec{u}}{u^2}\vec{u} \tag{7.46}$$

The transformations of the parallel component and the perpendicular component has been derived earlier. We note it again,

$$\vec{x}'_\parallel = \gamma(\vec{x}_\parallel - \vec{u}t) \tag{7.47}$$

$$\vec{x}'_\perp = \vec{x}_\perp \tag{7.48}$$

$$t' = \gamma\left(t - \frac{\vec{u} \cdot \vec{x}_\parallel}{c^2}\right) \tag{7.49}$$

and,

$$\begin{aligned} \vec{x}' &= \vec{x}'_\parallel + \vec{x}'_\perp & (7.50) \\ &= \gamma(\vec{x}_\parallel - \vec{u}t) + \vec{x}_\perp & (7.51) \\ &= \gamma\left(\frac{\vec{x}\cdot\vec{u}}{u^2}\vec{u} - \vec{u}t\right) + \vec{x} - \frac{\vec{x}\cdot\vec{u}}{u^2}\vec{u} & (7.52) \\ &= \vec{x} + (\gamma - 1)\frac{\vec{x}\cdot\vec{u}}{u^2}\vec{u} - \gamma\vec{u}t & (7.53) \end{aligned}$$

Since $\vec{x}_\perp \cdot \vec{u} = 0$, we can also write $\vec{x}_\parallel \cdot \vec{u} = \vec{x} \cdot \vec{u}$, so

$$t' = \gamma\left(t - \frac{\vec{u} \cdot \vec{x}}{c^2}\right) \tag{7.54}$$

Special Relativity 59

7.2 Basic consequences of special relativity

7.2.1 Time dilation

We recall the 2 inertial frames setup where S'-frame is moving at speed u with respect to the x-axis of the S-frame. Let a clock be in the S'-frame.[5]

- Let event 1 be the second-hand of the clock at the 12 o'clock-mark.
- Let event 2 be the second-hand of the clock at the 1 o'clock-mark.

Recall the inverse transformation: $t = \gamma \left(t' + \frac{u}{c^2}x'\right)$ and we want to find a relationship between time intervals as measured from each frame.

$$t_1 = \gamma \left(t'_1 + \frac{u}{c^2}x'_1\right) \text{ and } t_2 = \gamma \left(t'_2 + \frac{u}{c^2}x'_2\right) \tag{7.55}$$

Take difference, $t_2 - t_1 = \gamma \left(t'_2 - t'_1 + \frac{u}{c^2}(x'_2 - x'_1)\right) \tag{7.56}$

$$\Delta t = \gamma \left(\Delta t' + \frac{u}{c^2}\Delta x'\right) \tag{7.57}$$

| since the clock is in the same x' coordinate, $\Delta x' = 0$

$$\boxed{\Delta t = \gamma \Delta t'} \tag{7.58}$$

Thus if $\Delta t' = $ a 5 second interval (measured in S'-frame), then $\Delta t > 5$ seconds as measured in S-frame. In other words, if there is a similar clock in S-frame and both clocks are set to start ticking from the 12 o'clock-mark together, when the S'-clock is ticking at the 1 o'clock-mark, the S-clock is seen to be ticking at the γ-mark.[6]

7.2.2 Length contraction

Let a stick be on the x'-axis in the S'-frame.

- Let event 1 be the left end of the stick.
- Let event 2 be the right end of the stick.

Recall the inverse Lorentz transformation: $x = \gamma(x' + ut')$ and we want to find a relationship between spatial intervals as measured from each frame.

$$x_1 = \gamma(x'_1 + ut'_1) \text{ and } x_2 = \gamma(x'_2 + ut'_2) \tag{7.59}$$

Take difference, $x_2 - x_1 = \gamma(x'_2 - x'_1 + u(t'_2 - t'_1)) \tag{7.60}$

$$\Delta x = \gamma(\Delta x' + u\Delta t') \tag{7.61}$$

| seeing the stick means both ends are observed simultaneously $\underbrace{}_{\Delta t = 0}$

| then recall $\Delta t = \gamma \left(\Delta t' + \frac{u}{c^2}\Delta x'\right)$ so $\Delta t' = -\frac{u}{c^2}\Delta x'$

$$\Delta x = \gamma \left(\Delta x' - \frac{u^2}{c^2}\Delta x'\right) \tag{7.62}$$

---

[5] or S'-frame is comoving with a clock that is moving with speed u along the x-axis of the S-frame
[6] Assuming $\gamma < 12$ for simplicity.

$$\boxed{\Delta x = \frac{\Delta x'}{\gamma}} \quad \left| \text{ recall } \gamma = \frac{1}{\sqrt{1 - \frac{u^2}{c^2}}} \right. \tag{7.63}$$

So, $\Delta x'$ is the length of the stick measured in S'-frame where the stick has no relative motion and Δx is the length of the stick measured in S-frame and $\Delta x < \Delta x'$. This is due to relative simultaneity: we require both ends of the stick to be observed simultaneously in S-frame but actually these 2 events are not simultaneous in S'-frame. The opposite case is true also. This fact can be illustrated clearly on the Minkowski spacetime diagram which we will discuss later.

7.2.3 Velocity addition

We have already derived the velocity addition formula which is equation (7.35): $\boxed{v' = \dfrac{v - u}{1 - \frac{uv}{c^2}}}$ but let's state the situation more precisely.

Basically, there are 3 inertial frames: S, S' and S''. S'-frame is moving at speed u relative to S-frame and S''-frame is moving at speed v relative to S-frame. Then v' is the speed of S''-frame relative to S'-frame.

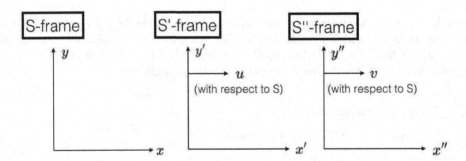

FIGURE 7.2
A 3 inertial frame setup.

We shall go further and relate the various γ factors: $\gamma(v') = \dfrac{1}{\sqrt{1 - \frac{v'^2}{c^2}}}$, $\gamma(u) = \dfrac{1}{\sqrt{1 - \frac{u^2}{c^2}}}$ and $\gamma(v) = \dfrac{1}{\sqrt{1 - \frac{v^2}{c^2}}}$,

$$\gamma(v') = \frac{1}{\sqrt{1 - \frac{v'^2}{c^2}}} \tag{7.64}$$

$$\gamma(v') = \left(1 - \frac{1}{c^2}\left(\frac{v-u}{1 - \frac{uv}{c^2}}\right)^2\right)^{-1/2} \tag{7.65}$$

$$\gamma(v') = \left(\frac{\left(1 - \frac{uv}{c^2}\right)^2 - \frac{(v-u)^2}{c^2}}{\left(1 - \frac{uv}{c^2}\right)^2}\right)^{-1/2} \tag{7.66}$$

Special Relativity

| expand the numerator

$$\gamma(v') = \left(1 - \frac{2uv}{c^2} + \frac{u^2v^2}{c^4} - \frac{v^2 - 2uv + u^2}{c^2}\right)^{-1/2} \left(1 - \frac{uv}{c^2}\right) \quad (7.67)$$

$$\gamma(v') = \left(1 - \frac{v^2}{c^2} - \frac{u^2}{c^2} + \frac{u^2v^2}{c^4}\right)^{-1/2} \left(1 - \frac{uv}{c^2}\right) \quad (7.68)$$

$$\gamma(v') = \left(\left(1 - \frac{u^2}{c^2}\right)\left(1 - \frac{v^2}{c^2}\right)\right)^{-1/2} \left(1 - \frac{uv}{c^2}\right) \quad (7.69)$$

$$\boxed{\gamma(v') = \gamma(u)\gamma(v)\left(1 - \frac{uv}{c^2}\right)} \quad (7.70)$$

We will need this identity whenever we have 3 frames and need to change perspectives between the 3 frames.

7.3 Further consequences (Minkowski spacetime diagram, invariant interval and causality)

7.3.1 Minkowski spacetime diagram

The Minkowski spacetime diagram is one way of graphically representing both S and S'-frames together so that by referencing the S-frame axes, we get the coordinates in S-frame and by referencing the S'-frame axes, we get the coordinates in S'-frame.

We will construct S-frame in the usual way where the ct vs x axis are 90 degrees to each other. The S'-frame shall be constructed based on Lorentz transformations.

- By definition, ct' axis means $x' = 0$,

$$x' = \gamma(x - ut) \xrightarrow{x'=0} 0 = x - ut \quad (7.71)$$

$$ct = \frac{c}{u}x \quad (7.72)$$

- By definition, x' axis means $t' = 0$,

$$t' = \gamma\left(t - \frac{ux}{c^2}\right) \xrightarrow{t'=0} 0 = t - \frac{ux}{c^2} \quad (7.73)$$

$$ct = \frac{u}{c}x \quad (7.74)$$

Note that the spacings are not the same on the axes of different frames: take coordinate $(x', ct') = (1, 0)$ and using Lorentz transformations

$$1 = \gamma(x - ut) \text{ and } 0 = t - \frac{ux}{c^2} \implies x = \gamma \text{ and } t = \frac{u}{c}\gamma \quad (7.75)$$

7.3.2 Invariant interval

Consider this strange "Pythagoras theorem" for "length" in S'-frame

$$-c^2t'^2 + x'^2 = -c^2\gamma^2\left(t - \frac{ux}{c^2}\right)^2 + \gamma^2(x - ut)^2 \quad (7.76)$$

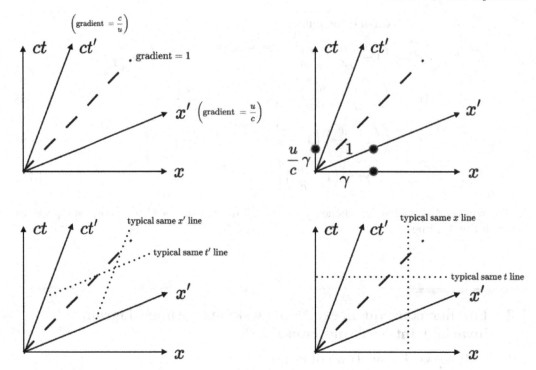

FIGURE 7.3
Summary of the main features in Minkowski spacetime diagram.

$$-c^2 t'^2 + x'^2 = \gamma^2 \left(-c^2 t^2 + \frac{2tuxc^2}{c^2} - \frac{u^2 x^2}{c^2} + x^2 - 2xut + u^2 t^2 \right) \tag{7.77}$$

$$-c^2 t'^2 + x'^2 = \gamma^2 \left(-c^2 t^2 + x^2 - \frac{u^2 x^2}{c^2} + u^2 t^2 \right) \tag{7.78}$$

$$-c^2 t'^2 + x'^2 = \frac{1}{1 - \frac{u^2}{c^2}} \left(-c^2 t^2 \left(1 - \frac{u^2}{c^2}\right) + x^2 \left(1 - \frac{u^2}{c^2}\right) \right) \tag{7.79}$$

$$-c^2 t'^2 + x'^2 = -c^2 t^2 + x^2 \tag{7.80}$$

so this "length" is the same in S-frame and therefore this "length" is called the invariant interval because it is the same (numerical value) regardless of the inertial frame.[7]

If we choose a certain inertial frame where a clock is stationary, the time measured (in this comoving frame) is called the "proper time" and is denoted by τ. As the clock is always at the origin in this frame, $x^2 = 0$ and so the invariant interval in this frame is $-c^2 \tau^2$. We can write the following forms:

- $-c^2 \tau^2 = -c^2 t^2 + x^2$

- Full 4D: $-c^2 \tau^2 = -c^2 t^2 + x^2 + y^2 + z^2$

- Define infinitesimal invariant interval ds: $(ds)^2 = -c^2 (d\tau)^2 = -c^2 (dt)^2 + (dx)^2 + (dy)^2 + (dz)^2$

- Conventional writing: $\boxed{ds^2 = -c^2 d\tau^2 = -c^2 dt^2 + dx^2 + dy^2 + dz^2}$ [8]

---

[7] Eventually, we will call this a Lorentz scalar which are objects with the same value in every inertial frame.
[8] It is also called the metric, which is the same name as the metric tensor introduced later.

Special Relativity

Since there are 3 positive terms and 1 negative term in the invariant interval, it can be positive, zero or negative. We illustrate this on the Minkowski diagram.

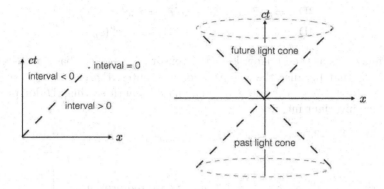

FIGURE 7.4
3 separate regions in the Minkowski spacetime diagram.

- There should be a z-axis also but I can't draw it. Thus the light cones should really be hypercones. The 3 regions are named as follows:

    - For interval < 0, it is called the timelike region. This is the region where particles with mass move because particles with mass can only move slower than c.
    - For interval $= 0$, it is called the lightlike region. This is the region where photons travel because photons only travel with speed c.
    - For interval > 0, it is called the spacelike region. This is the region where particles move faster than c.

7.4 Transition into tensor calculus

7.4.1 Minkowski metric, 4-vectors

We will now carry out a more formal treatment so that we can handle 4D vectors (in Minkowski space). This is an "upgrade" of the treatment of 3D vectors (previously denoted as $\hat{x}, \hat{y}, \hat{z}$).

$$3D \Rightarrow \vec{x} = \sum_i x^i \hat{e}_i \text{ where } \hat{e}_i = \hat{e}_x, \hat{e}_y, \hat{e}_z \text{ are the basis vectors} \tag{7.81}$$

$$4D \Rightarrow \bar{x} = \sum_\mu x^\mu \hat{e}_\mu \text{ where } \hat{e}_\mu = \hat{e}_{ct}, \hat{e}_x, \hat{e}_y, \hat{e}_z \text{ are the basis vectors} \tag{7.82}$$

Now we introduce the Einstein summation convention: whenever 2 identical indices appear (one in the superscript, one in the subscript), a summation is implied. Greek indices run from 0 to 3 and Latin indices run from 1 to 3. Indices $0 = ct$, $1 = x$, $2 = y$ and $3 = z$.

Now consider the dot product of 3D and 4D vectors with itself which gives the squared length:

$$3D \Rightarrow \vec{x} \cdot \vec{x} = x^i \hat{e}_i \cdot x^j \hat{e}_j = x^i x^j (\hat{e}_i \cdot \hat{e}_j) \tag{7.83}$$

$$4D \Rightarrow \bar{x} \cdot \bar{x} = x^\mu \hat{e}_\mu \cdot x^\nu \hat{e}_\nu = x^\mu x^\nu (\hat{e}_\mu \cdot \hat{e}_\nu) \tag{7.84}$$

We shall define $\hat{e}_\mu \cdot \hat{e}_\nu$ as the Minkowski metric tensor $\eta_{\mu\nu}$ since the basis vectors characterise the space. Note that because the dot product commutes, therefore $\eta_{\mu\nu} = \eta_{\nu\mu}$. Note that the 4D dot product is supposed to give the squared length so this 4D dot product should be equal to the invariant interval.

$$-c^2 t^2 + x^2 + y^2 + z^2 = \bar{x} \cdot \bar{x} = x^\mu x^\nu (\hat{e}_\mu \cdot \hat{e}_\nu) = x^\mu x^\nu \eta_{\mu\nu} \tag{7.85}$$

with $x^\mu, x^\nu = (x^0, x^1, x^2, x^3) = (ct, x, y, z)$ we require

$$\boxed{\eta_{\mu\nu} = \begin{pmatrix} -1 & & & \\ & 1 & & \\ & & 1 & \\ & & & 1 \end{pmatrix}} \tag{7.86}$$

We can take another perspective where the dot product is formed by a vector from a space with another vector from the "dual" space.[9]

Define the position 4-vector with an upper index:
$$x^\mu = (x^0, x^1, x^2, x^3) \tag{7.87}$$
$$x^\mu = (ct, x, y, z) \tag{7.88}$$

Define the position 4-vector with a lower index:
$$x_\mu = \eta_{\mu\nu} x^\nu \tag{7.89}$$

$$x_\mu = \begin{pmatrix} -1 & & & \\ & 1 & & \\ & & 1 & \\ & & & 1 \end{pmatrix} \begin{pmatrix} ct \\ x \\ y \\ z \end{pmatrix}$$

$$x_\mu = \begin{pmatrix} -ct \\ x \\ y \\ z \end{pmatrix} \tag{7.90}$$

$$x_\mu = (-ct, x, y, z) \tag{7.91}$$

The dot product: $x_\mu x^\mu = \begin{pmatrix} -ct & x & y & z \end{pmatrix} \begin{pmatrix} ct \\ x \\ y \\ z \end{pmatrix}$

$$x_\mu x^\mu = -c^2 t^2 + x^2 + y^2 + z^2 \tag{7.92}$$

and the metric tensor maps an upper index vector to a lower index vector (in dual space). This is casually described as the metric tensor "lowering an index".

To map from a lower vector to an upper index vector (or "raising an index"), we need the inverse metric tensor.

$$x^\mu = \eta^{\mu\nu} x_\nu = \eta^{\mu\nu} \eta_{\nu\sigma} x^\sigma = \delta^\mu_\sigma x^\sigma \tag{7.93}$$

---

[9]Recall from maths that inner product is always formed by 2 objects: one from vector space, the other from dual space. This is like in quantum mechanics where the dot product is formed between a vector in ket-space and another vector in bra-space and the 2 spaces are Hermitian conjugates of each other.e.

Special Relativity

which means $\eta^{\mu\nu}$ is the inverse of $\eta_{\mu\nu}$.

Explicitly, $\boxed{\eta^{\mu\nu} = \begin{pmatrix} -1 & & & \\ & 1 & & \\ & & 1 & \\ & & & 1 \end{pmatrix}}$ since $\delta^\mu_\sigma = \begin{pmatrix} 1 & & & \\ & 1 & & \\ & & 1 & \\ & & & 1 \end{pmatrix}$.

7.4.2 Momentum 4-vector and Einstein relation

We want to define a velocity 4-vector and proceed to construct a momentum 4-vector from there. A quantity is a 4-vector if it transforms like a 4-vector under Lorentz transformations.[10]

$$\text{Define the contravariant velocity 4-vector: } \boxed{u^\mu = \frac{dx^\mu}{d\tau}} \quad (7.94)$$

This guess is made by requiring a time derivative on the position 4-vector and choosing the time to be the proper time because the numerator is already a 4-vector and so we do not want the denominator to transform also. Thus we can

$$\text{Define the contravariant momentum 4-vector: } \boxed{p^\mu = mu^\mu = m\frac{dx^\mu}{d\tau}} \quad (7.95)$$

where m is called the rest mass which is the mass measured in the comoving frame of the particle.

Now we need to check that the objects are indeed 4-vectors. First we work out the time component (or 0-component),

$$p^0 = m\frac{dx^0}{d\tau} = m\frac{cdt}{d\tau} \quad (7.96)$$

$$\mid \text{ from time dilation, } dt = \gamma(u_1)d\tau, \; \gamma(u_1) = \frac{1}{\sqrt{1-\frac{u_1^2}{c^2}}}$$

$$p^0 = mc\gamma(u_1) \quad (7.97)$$

where u_1 is the relative velocity of the particle with respect to the S-frame. The comoving frame of the particle is called the S''-frame.

We expect a transformation of the form for the space component:

$$x' = \gamma(x - ut) = \gamma\left(x - \frac{u}{c}(ct)\right) \quad (7.98)$$

$$\Rightarrow (\text{"space component"})' = \gamma\left(\text{"space component"} - \frac{u}{c}(\text{"time component"})\right) \quad (7.99)$$

So we check from S'-frame which is moving at speed u with respect to S-frame,

$$p'^1 = m\frac{dx'^1}{d\tau} \quad (7.100)$$

$$p'^1 = m\frac{dx'}{d\tau} \quad (7.101)$$

$$\mid \text{ recall } dx' = \gamma(u)(dx - udt)$$

---

[10] Define any vector or tensor you like but it must Lorentz transform correctly.

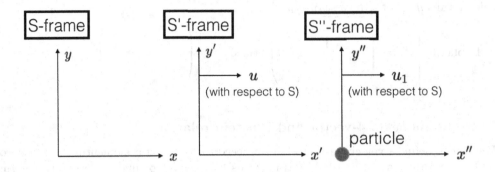

FIGURE 7.5
3 inertial frame setup for deriving the Lorentz transformations of the momentum 4-vector.

$$p'^1 = m\gamma(u)\left(\frac{dx}{d\tau} - u\frac{dt}{d\tau}\right) \qquad (7.102)$$

| recall time dilation $dt = \gamma(u_1)d\tau$

$$p'^1 = m\gamma(u)\left(\frac{p^1}{m} - u\gamma(u_1)\right) \qquad (7.103)$$

$$p'^1 = \gamma(u)\left(p^1 - \frac{u}{c}(mc\gamma(u_1))\right) \qquad (7.104)$$

$$p'^1 = \gamma(u)\left(p^1 - \frac{u}{c}p^0\right) \qquad (7.105)$$

Indeed, p^1 transforms like the first space component. Now we check the time component which is expected to transform as

$$t' = \gamma\left(t - \frac{u}{c^2}x\right) \qquad (7.106)$$

$$\Rightarrow ct' = \gamma\left(ct - \frac{u}{c}x\right) \qquad (7.107)$$

$$\Rightarrow (\text{"time component"})' = \gamma\left(\text{"time component"} - \frac{u}{c}(\text{"space component"})\right) \qquad (7.108)$$

$$\text{So, } p'^0 \stackrel{?}{=} \gamma(u)\left(p^0 - \frac{u}{c}p^1\right) \qquad (7.109)$$

$$\text{The LHS} = p'^0 = mc\gamma(u'_1) \qquad (7.110)$$

with $\gamma(u_1) = \dfrac{1}{\sqrt{1 - \frac{u_1'^2}{c^2}}}$ where u'_1 is the velocity of the particle (S''-frame) with respect to the S'-frame. Velocity addition says that $u'_1 = \dfrac{u_1 - u}{1 - \frac{u_1 u}{c^2}}$.

$$\text{The RHS} = \gamma(u)\left(p^0 - \frac{u}{c}p^1\right) \qquad (7.111)$$

$$= \gamma(u)\left(mc\gamma(u_1) - \frac{u}{c}m\frac{dx}{d\tau}\right) \qquad (7.112)$$

| recall $dt = \gamma(u_1)d\tau$ and $\dfrac{dx}{dt} = u_1$

$$= \gamma(u)\left(mc\gamma(u_1) - \frac{u}{c}mu_1\gamma(u_1)\right) \qquad (7.113)$$

Special Relativity

$$= mc\gamma(u)\gamma(u_1)\left(1 - \frac{uu_1}{c^2}\right) \quad (7.114)$$

| recall the identity, eq (7.70): $\gamma(u)\gamma(u_1)\left(1 - \frac{uu_1}{c^2}\right) = \gamma(u_1')$

$$= mc\gamma(u_1') \quad (7.115)$$

Thus indeed p^0 transforms as a time component. We rewrite the upper index momentum 4-vector as,

$$p^\mu = (p^0, p^1, p^2, p^3) = \left(\frac{E}{c}, p_x, p_y, p_z\right) = (mc\gamma(u_1), mu_1\gamma(u_1), 0, 0) \quad (7.116)$$

We shall check that $E = mc^2\gamma(u_1)$ is indeed the total energy.

$$E = mc^2\gamma(u_1) = mc^2\left(1 - \frac{u_1^2}{c^2}\right)^{-1/2} \quad (7.117)$$

| assume the particle moves slowly, so $u_1 << c$

$$E = mc^2\left(1 + \frac{1}{2}\frac{u_1^2}{c^2} + \cdots\right) \quad (7.118)$$

$$E = \underbrace{mc^2}_{\text{rest mass energy}} + \underbrace{\frac{1}{2}mu_1^2}_{\text{Newtonian KE}} + \underbrace{\cdots}_{\text{relativistic corrections}} \quad (7.119)$$

Recall that for the position 4-vector, we carried out the dot product to find its (squared) "length" and that quantity is a Lorentz invariant quantity (Lorentz scalar). We do the same for the momentum 4-vector,

$$\text{lower index momentum 4-vector:} \quad p_\mu = \eta_{\mu\nu}p^\nu \quad (7.120)$$

$$p_\mu = \left(-\frac{E}{c}, p_x, p_y, p_z\right) \quad (7.121)$$

$$\text{(squared) "length":} \quad \bar{p} \cdot \bar{p} = p_\mu p^\mu = -\frac{E^2}{c^2} + p_x^2 + p_y^2 + p_z^2 = -\frac{E^2}{c^2} + \vec{p}^2 \quad (7.122)$$

We can also form $p^\mu p_\mu$ by using $p^\mu = (mc\gamma(u_1), mu_1\gamma(u_1), 0, 0)$

$$p^\mu p_\mu = -m^2c^2\gamma(u_1)^2 + m^2u_1^2\gamma(u_1)^2 = -m^2c^2\gamma(u_1)^2\left(1 - \frac{u_1^2}{c^2}\right) = -m^2c^2 \quad (7.123)$$

and we equate the two expressions of $p^\mu p_\mu$,

$$-m^2c^2 = -\frac{E^2}{c^2} + \vec{p}^2 \quad (7.124)$$

$$\boxed{E^2 = \vec{p}^2c^2 + m^2c^4} \quad (7.125)$$

which is the famous Einstein relation. From the derivation, now you know that E and $\vec{p}$ of any frame can be used and they should be taken from the same frame.

7.4.3 Derivative 4-vectors and d'Alembert's operator

Intuitively, we define the derivative 4-vector as

$$\frac{\partial}{\partial x^\mu} = \left(\frac{1}{c}\frac{\partial}{\partial t}, \frac{\partial}{\partial x}, \frac{\partial}{\partial y}, \frac{\partial}{\partial z}\right) \quad \text{and} \quad \frac{\partial}{\partial x_\mu} = \left(-\frac{1}{c}\frac{\partial}{\partial t}, \frac{\partial}{\partial x}, \frac{\partial}{\partial y}, \frac{\partial}{\partial z}\right) \quad (7.126)$$

We need to check that they are really 4-vectors by checking their transformations.

$$\frac{\partial}{\partial x'^0} = \frac{1}{c}\frac{\partial}{\partial t'} = \frac{\partial x^\mu}{\partial ct'}\frac{\partial}{\partial x^\mu} \tag{7.127}$$

$$\frac{\partial}{\partial ct'} = \frac{\partial ct}{\partial ct'}\frac{\partial}{\partial ct} + \frac{\partial x}{\partial ct'}\frac{\partial}{\partial x} + 0 + 0 \tag{7.128}$$

| recall Lorentz transformations $ct = \gamma\left(ct' + \frac{u}{c}x'\right)$ and $x = \gamma(x' + ut')$

$$\frac{\partial}{\partial ct'} = \gamma\frac{\partial}{\partial ct} + \gamma\frac{u}{c}\frac{\partial}{\partial x} \tag{7.129}$$

so if we recall the Lorentz transformation $ct' = \gamma\left(ct - \frac{u}{c}x\right)$, we can deduce that $\frac{\partial}{\partial ct'} = \frac{\partial}{\partial x^0}$ transforms like the time component of a lower index 4-vector. To confirm, we check further,

$$\frac{\partial}{\partial x'^1} = \frac{\partial}{\partial x'} = \frac{\partial ct}{\partial x'}\frac{\partial}{\partial ct} + \frac{\partial x}{\partial x'}\frac{\partial}{\partial x} + 0 + 0 \tag{7.130}$$

| recall the Lorentz transformations $ct = \gamma\left(ct' + \frac{u}{c}x'\right)$ and $x = \gamma(x' + ut')$

$$\frac{\partial}{\partial x'} = \gamma\frac{u}{c}\frac{\partial}{\partial ct} + \gamma\frac{\partial}{\partial x} \tag{7.131}$$

$$\frac{\partial}{\partial x'} = \gamma\left(\frac{\partial}{\partial x} + \frac{u}{c}\frac{\partial}{\partial ct}\right) \tag{7.132}$$

so if we recall the Lorentz transformation $x' = \gamma(x - ut)$, we again realise that the time component behaves with a "wrong" sign which means that it is the time component of a lower index 4-vector, thus $\frac{\partial}{\partial x^\mu}$ is a lower index 4-vector.

We can immediately check for $\frac{\partial}{\partial x_\mu}$,

$$\frac{\partial}{\partial x'_0} = -\frac{1}{c}\frac{\partial}{\partial t'} = \gamma\left(\left(-\frac{1}{c}\frac{\partial}{\partial t} - \frac{u}{c}\frac{\partial}{\partial x}\right)\right) \tag{7.133}$$

$$\frac{\partial}{\partial x'_1} = \frac{\partial}{\partial x'} = \gamma\left(\frac{\partial}{\partial x} - \frac{u}{c}\left(-\frac{\partial}{\partial ct}\right)\right) \tag{7.134}$$

which means $\frac{\partial}{\partial x_\mu}$ is behaving like an upper index 4-vector! Hence we denote:[11]

$$\boxed{\partial_\mu = \frac{\partial}{\partial x^\mu} = \left(\frac{\partial}{\partial x^0}, \frac{\partial}{\partial x^1}, \frac{\partial}{\partial x^2}, \frac{\partial}{\partial x^3}\right) = \left(\frac{1}{c}\frac{\partial}{\partial t}, \frac{\partial}{\partial x}, \frac{\partial}{\partial y}, \frac{\partial}{\partial z}\right)} \tag{7.135}$$

$$\boxed{\partial^\mu = \frac{\partial}{\partial x_\mu} = \left(\frac{\partial}{\partial x_0}, \frac{\partial}{\partial x_1}, \frac{\partial}{\partial x_2}, \frac{\partial}{\partial x_3}\right) = \left(-\frac{1}{c}\frac{\partial}{\partial t}, \frac{\partial}{\partial x}, \frac{\partial}{\partial y}, \frac{\partial}{\partial z}\right)} \tag{7.136}$$

We shall now take the dot product of these 2 vectors. Recall that it is supposed to give us a quantity that is Lorentz invariant (Lorentz scalar).

$$\boxed{\partial^\mu \partial_\mu = -\frac{1}{c^2}\frac{\partial^2}{\partial t^2} + \frac{\partial^2}{\partial x^2} + \frac{\partial^2}{\partial y^2} + \frac{\partial^2}{\partial z^2}} \tag{7.137}$$

$$\boxed{\Box = -\frac{1}{c^2}\frac{\partial^2}{\partial t^2} + \vec{\nabla}^2} \tag{7.138}$$

where $\partial^\mu \partial_\mu = \Box$ is called the 4D Laplacian or d'Alembertian operator and $\vec{\nabla}^2$ is the 3D Laplacian operator.[12]

---

[11] The partial derivative is the only weird object where it looks like an "upper index" $\frac{\partial}{\partial x^\mu}$ but has the transformation properties of a "lower index" object, so we label as ∂_μ.

[12] In some EM books, the d'Alembertian has the symbol $\Box^2 = \partial^\mu \partial_\mu$.

Special Relativity

7.4.4 Relativistic mechanics

Newton's first law is built into the first axiom of special relativity in the sense that in inertial frames, there is no net force due to relative motion and so different inertial frames would experience the same physics.

For Newton's second law, the most intuitive way to fit it into special relativity is to modify the momentum to relativistic momentum.

$$\vec{F} = \frac{d\vec{p}}{dt} = \frac{d}{dt}\left(m\vec{v}\gamma(v)\right) \quad \text{(where } v \text{ is the speed of the particle)} \tag{7.139}$$

This relation is for S-frame where the particle is seen to be moving at speed v. How about in S'-frame which is moving at speed u with respect to S-frame along the x axis? We have to transform both numerator and denominator! For simplicity, we shall take S'-frame to be the (instantaneous) comoving frame of the particle, i.e. $v = u$. For the y component,

$$F'_y = \frac{dp'_y}{dt'} = \frac{dp_y}{\gamma(u)\left(dt - \frac{u}{c^2}dx\right)} = \frac{\frac{dp_y}{dt}}{\gamma(u)\left(1 - \frac{u}{c^2}\frac{dx}{dt}\right)} \tag{7.140}$$

$$\quad \text{where } \frac{dx}{dt} = u$$

$$F'_y = \frac{F_y}{\gamma(u)\left(1 - \frac{u^2}{c^2}\right)} = \gamma(u)F_y \tag{7.141}$$

Similarly for the z component

$$F'_z = \gamma(u)F_z \tag{7.142}$$

Now for the x component,

$$F'_x = \frac{dp'_x}{dt'} = \frac{\gamma(u)\left(dp_x - \frac{u}{c}dp^0\right)}{\gamma(u)\left(dt - \frac{u}{c^2}dx\right)} = \frac{\frac{dp_x}{dt} - \frac{u}{c}\frac{dp^0}{dt}}{1 - \frac{u}{c^2}\frac{dx}{dt}} \tag{7.143}$$

$$= \frac{F_x - \frac{u}{c^2}\frac{dE}{dt}}{1 - \frac{u^2}{c^2}} \tag{7.144}$$

$$\quad \text{where } \frac{dE}{dt} = \frac{d}{dt}mc^2\gamma(u) = mc^2\frac{\frac{2u}{c^2}}{2\left(1 - \frac{u^2}{c^2}\right)^{3/2}}\frac{du}{dt} = u\frac{d}{dt}m u\gamma(u) = uF_x$$

$$F'_x = F_x \tag{7.145}$$

Thus the ordinary force transforms awkwardly. We try to define another force quantity which is a 4-vector. We could take the clue from the definition of 4-velocity and define the 4-force (or Minkowski force) as

$$\boxed{K^\mu = \frac{dp^\mu}{d\tau}} \tag{7.146}$$

where p^μ is the 4-momentum and τ is the proper time. The spatial components are

$$\vec{K} = \left(\frac{dt}{d\tau}\right)\frac{d\vec{p}}{dt} = \gamma(u)\vec{F} \quad \text{(where } \vec{F} \text{ is the ordinary force)} \tag{7.147}$$

The time component is

$$K^0 = \frac{dp^0}{d\tau} = \frac{1}{c}\frac{dE}{d\tau} \qquad (7.148)$$

where $\frac{dE}{d\tau}$ can be interpreted as the rate of energy increase with respect to proper time or proper power.

There is nothing wrong with using $\vec{F} = \frac{d\vec{p}}{dt}$ where $\vec{p}$ is the relativistic momentum except for awkwardness in transformation. Therefore we shall use the covariant Minkowski force K^μ for the sake of convenience in doing Lorentz transformations.

8

Electrodynamics Recast into a Manifestly Lorentz Covariant Form

8.1 Just for motivation: Magnetism as a relativistic phenomena

The objective for this chapter is to make it clear that electrodynamics is meant to be described in Minkowski space. The $\vec{E}$ and $\vec{B}$ fields are just 2 sides of the same coin. This means that they are 6 components of a single object called the field strength tensor $F^{\mu\nu}$. The $\vec{E}$ and $\vec{B}$ fields shouldn't really be seen as 2 different fields. The rewriting of electrodynamics in the language of special relativity does not really generalise anything. It shows this unification of $\vec{E}$ and $\vec{B}$ clearly.

So in this section, we want to deduce magnetism from the knowledge of electrostatics and special relativity to motivate the intimate connection between electrodynamics and special relativity.

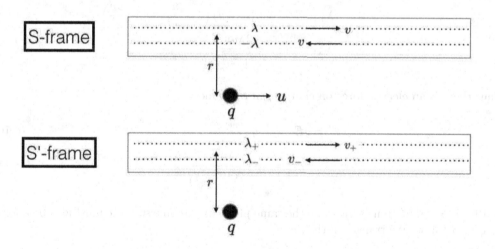

Consider, in S-frame, we see a wire and inside the wire, we see a string of positive charges moving to the right with speed v. We assume that a linear charge density λ can be defined for it. In the same wire, superimposed on the positive string of charges is a negative one, with $-\lambda$ moving to the left with speed v. The net current is to the right and given by

$$I = 2\lambda v \tag{8.1}$$

For a charge q, distance r outside the wire and moving to the right with speed $u(<v)$, there is no electrical force since the 2-line charges have no net charge.

We now go into S'-frame which is moving to the right with speed u with respect to S-frame. Using the velocity addition formula,

$$\text{speed of positive line charge in } S'\text{-frame: } v_+ = \frac{v-u}{1-\frac{vu}{c^2}} \tag{8.2}$$

$$\text{speed of negative line charge in } S'\text{-frame: } v_- = \frac{v+u}{1+\frac{vu}{c^2}} \tag{8.3}$$

The linear charge densities are seen to change by different amounts due to length contraction.

$$\text{For positive (in } S'\text{-frame): } \lambda_+ = \gamma(v_+)\lambda_0 \text{ with } \gamma(v_+) = \frac{1}{\sqrt{1-\frac{v_+^2}{c^2}}} \tag{8.4}$$

$$\text{For negative (in } S'\text{-frame): } \lambda_- = \gamma(v_-)\lambda_0 \text{ with } \gamma(v_-) = \frac{1}{\sqrt{1-\frac{v_-^2}{c^2}}} \tag{8.5}$$

where λ_0 is the charge density seen in the comoving frame of the charges,

$$\text{then, } \lambda = \gamma(v)\lambda_0 \text{ with } \gamma(v) = \frac{1}{\sqrt{1-\frac{v^2}{c^2}}} \tag{8.6}$$

and using the velocity addition identities, equation (7.70): $\gamma(v_+) = \gamma(v)\gamma(u)\left(1-\frac{uv}{c^2}\right)$ and $\gamma(v_-) = \gamma(v)\gamma(u)\left(1+\frac{uv}{c^2}\right)$, we can calculate the net charge seen in S'-frame

$$\lambda_{\text{net}} = \lambda_+ + \lambda_- = \lambda_0\left(\gamma(v_+) - \gamma(v_-)\right) \tag{8.7}$$

$$\lambda_{\text{net}} = \lambda_0\gamma(v)\gamma(u)\left(1-\frac{uv}{c^2}-1-\frac{uv}{c^2}\right) \tag{8.8}$$

$$\lambda_{\text{net}} = \frac{-2\lambda uv}{c^2\sqrt{1-\frac{u^2}{c^2}}} \tag{8.9}$$

Thus there is an electric force on charge q in S'-frame,

$$F' = qE = q\frac{\lambda_{\text{net}}}{2\pi\epsilon_0 r} \tag{8.10}$$

$$F' = -\frac{\lambda v}{\pi\epsilon_0 c^2 r}\frac{qu}{\sqrt{1-\frac{u^2}{c^2}}} \tag{8.11}$$

But both S and S'-frames must see the same physical outcome so there must also be a force on q in S-frame. We transform the force

$$F = \frac{1}{\gamma(u)}F' \quad \text{(refer to equation (7.141))} \tag{8.12}$$

$$F = -\frac{\lambda v}{\pi\epsilon_0 c^2}\frac{qu}{r} \tag{8.13}$$

$$\left| \text{ write } \frac{1}{c^2} = \epsilon_0\mu_0 \text{ and } 2\lambda v = I \right.$$

$$F = -qu\underbrace{\left(\frac{\mu_0 I}{2\pi r}\right)}_{\text{magnetic field in } S\text{-frame}} \tag{8.14}$$

So this force is the Lorentz force on a charged particle moving in a magnetic field.

8.2 Transformations of $\vec{E}$ and $\vec{B}$

To begin the rewriting of electrodynamics into Lorentz covariant form[1], we must first know how the $\vec{E}$ and $\vec{B}$ fields transform and realise that they do not form a 4-vector.

Consider a flat plate capacitor moving with speed u'' with respect to S-frame. Let the capacitor be in S''-frame, i.e. this S''-frame is the comoving frame of the capacitor.

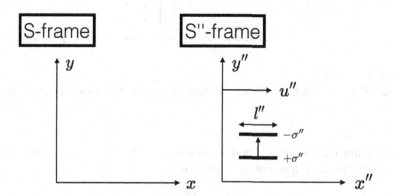

FIGURE 8.1
Capacitor in S''-frame and seen from S-frame. Capacitor plates are parallel to x'' axis.

In S''-frame, the capacitor has length l'' and width w and surface charge density σ''. Thus the electric field seen in S''-frame:

$$E_y'' = \frac{\sigma''}{\epsilon_0} \tag{8.15}$$

In S-frame, the length of the capacitor is measured with length contraction

$$l = \frac{1}{\gamma(u'')} l'' \quad \text{where} \quad \gamma(u'') = \frac{1}{\sqrt{1 - \frac{u''^2}{c^2}}} \tag{8.16}$$

Since total charge Q is invariant[2] and width w does not undergo length contraction as it is perpendicular to the direction of motion, the charge density measured in S-frame is $\sigma = \frac{Q}{lw} = \gamma(u'') \frac{Q}{l'' w} = \gamma(u'') \sigma''$.

The electric field measured in S-frame is[3]

$$E_y = \frac{\sigma}{\epsilon_0} = \gamma(u'') \frac{\sigma''}{\epsilon_0} = \gamma(u'') E_y'' \tag{8.17}$$

---

[1]Lorentz covariant means the expression has the same form in all inertial frames.
[2]I am taking this to be an empirical fact. It can be shown by Noether's theorem which will be discussed in section 10.3.
[3]We may suspect that the electric field may not be perpendicular to the plates but by symmetry considerations, any parallel component from the $+\sigma$ plate will be cancelled by an opposite parallel component from the $-\sigma$ plate.

Now consider the capacitor plates to be perpendicular to the x'' axis,

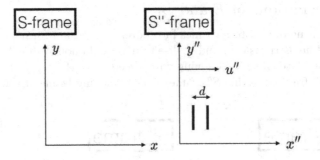

FIGURE 8.2
Capacitor in S''-frame and seen from S-frame. Capacitor plates are perpendicular to x'' axis.

So the capacitor plate spacing d undergoes length contraction but the electric field (which is now in the x direction) does not depend on d so,

$$E_x = E''_x \qquad (8.18)$$

We can now talk about the transformation of magnetic fields because in S-frame, 2 moving charged plates (we are back to the situation where the capacitor plates are parallel to the x'' axis) amounts to 2 surface currents.

In S-frame, the magnetic field points in the $+z$ direction between the plates using the right hand rule. Using Ampere's law, the magnetic field between the plates is

$$\text{In } S\text{-frame: } B_z = \mu_0 |K| = \mu_0 \sigma u'' \qquad (8.19)$$

To compare Lorentz transformed fields, we need to bring a 3rd inertial frame and we call it the S'-frame which is moving at speed u' with respect to S-frame.

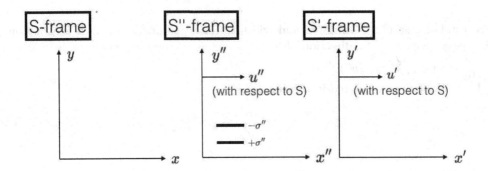

FIGURE 8.3
Capacitor in S'''-frame and seen from S-frame and from S'-frame. Capacitor plates are parallel to x'' axis.

From the addition of velocities, the relative velocity of S''-frame with respect to S'-frame is $u_1 = \dfrac{u'' - u'}{1 - \frac{u''u'}{c^2}}$.

$$\text{In } S'\text{-frame: } E'_y = \frac{\sigma'}{\epsilon_0} = \gamma(u_1)\frac{\sigma''}{\epsilon_0} = \gamma(u_1)E''_y \tag{8.20}$$

$$\text{In } S'\text{-frame: } B'_z = \mu_0 \sigma' u_1 = \mu_0 \gamma(u_1) \sigma'' u_1 \tag{8.21}$$

Finally, we need to relate E'_y to E_y and B_z and relate B'_z to E_y and B_z. Eliminating the "double primed" fields will give us the Lorentz transformation of the fields between S-frame and S'-frame.

$$E'_y = \gamma(u_1) E''_y \tag{8.22}$$

| recall the identity $\gamma(u_1) = \gamma(u'')\gamma(u')\left(1 - \dfrac{u''u'}{c^2}\right)$
| and recall equation (8.20): $\gamma(u'')E''_y = E_y$

$$E'_y = \gamma(u') E_y \left(1 - \frac{u''u'}{c^2}\right) \tag{8.23}$$

$$E'_y = \gamma(u') \left(E_y - \frac{u'}{c^2} u'' E_y\right) \tag{8.24}$$

$$E'_y = \gamma(u') \left(E_y - \frac{u'}{c^2} u'' \frac{\sigma}{\epsilon_0}\right) \tag{8.25}$$

| recall equation (8.19): $B_z = \mu_0 \sigma u''$

$$E'_y = \gamma(u') \left(E_y - \frac{u'}{c^2 \epsilon_0 \mu_0} B_z\right) \tag{8.26}$$

| recall $\dfrac{1}{c^2} = \epsilon_0 \mu_0$

$$E'_y = \gamma(u') \left(E_y - u' B_z\right) \tag{8.27}$$

Then for B'_z

$$B'_z = \mu_0 \gamma(u_1) \sigma'' u_1 \tag{8.28}$$

| recall identity $\gamma(u_1) = \gamma(u'')\gamma(u')\left(1 - \dfrac{u''u'}{c^2}\right)$ and $u_1 = \dfrac{u'' - u'}{1 - \frac{u''u'}{c^2}}$
| also use equation (8.15): $\sigma'' = E''_y \epsilon_0$

$$B'_z = \mu_0 \gamma(u'') \gamma(u') E''_y \epsilon_0 (u'' - u') \tag{8.29}$$

| recall equation (8.20): $\gamma(u'') E''_y = E_y$

$$B'_z = \epsilon_0 \mu_0 \gamma(u') E_y (u'' - u') \tag{8.30}$$

$$B'_z = \gamma(u') \left(\epsilon_0 \mu_0 E_y u'' - \epsilon_0 \mu_0 E_y u'\right) \tag{8.31}$$

| recall eq(8.17): $E_y = \dfrac{\sigma}{\epsilon_0}$ and eq(8.19): $B_z = \mu_0 \sigma u''$ and $\epsilon_0 \mu_0 = \dfrac{1}{c^2}$

$$B'_z = \gamma(u') \left(B_z - \frac{u'}{c^2} E_y\right) \tag{8.32}$$

We can repeat with the capacitor plates set parallel to the xy plane and get

$$E'_z = \gamma(u')(E_z + u'B_y) \tag{8.33}$$

$$B'_y = \gamma(u')\left(B_y + \frac{u'}{c^2}E_z\right) \tag{8.34}$$

To get the transformation for B_x, we imagine a solenoid in S''-frame, with its axis aligned to the x'' axis.

$$B''_x = \mu_0 n'' I'' \tag{8.35}$$

$$\quad | \quad \text{in } S\text{-frame, time dilation gives } I = \frac{1}{\gamma(u'')}I''$$

$$\quad | \quad \text{in } S\text{-frame, length contraction gives } n = \gamma(u'')n''$$

$$B''_x = \mu_0 n I \tag{8.36}$$

$$B''_x = B_x \tag{8.37}$$

So for $\vec{E}$ and $\vec{B}$, the component parallel to motion is unaffected and this is unusual because so far we do not have any quantity transforming in this way.

Finally, we collect the whole set of transformations,

$$\boxed{\begin{array}{lll} E'_x = E_x & E'_y = \gamma(u')(E_y - u'B_z) & E'_z = \gamma(u')(E_z + u'B_y) \\ B'_x = B_x & B'_y = \gamma(u')\left(B_y + \frac{u'}{c^2}E_z\right) & B'_z = \gamma(u')\left(B_z - \frac{u'}{c^2}E_y\right) \end{array}} \tag{8.38}$$

This set of transformation is nowhere similar to the transformation of a 4-vector. This also gives a strong hint that $\vec{E}$ and $\vec{B}$ are not different fields as they transform into each other.

8.3 Rewriting electrodynamics or unification of $\vec{E}$ and $\vec{B}$

8.3.1 Maxwell's equations

Transformations of $\vec{E}$ and $\vec{B}$ are not like that of a 4-vector but they are that of an antisymmetric 2-tensor.[4] An antisymmetric (4D) 2-tensor has 6 elements and there are exactly 6 field components in $\vec{E}$ and $\vec{B}$. Thus the conclusion is that, there is only one unified field in electrodynamics and it is the 2-tensor field strength denoted as $F^{\mu\nu}$.[5]

We take on a matrix way of writing:

Position 4-vector:

$$\begin{array}{l} ct' = \gamma(u)\left(ct - \frac{u}{c}x\right) \\ x' = \gamma(u)(x - ut) \\ y' = y \\ z' = z \end{array} \implies \begin{pmatrix} ct' \\ x' \\ y' \\ z' \end{pmatrix} = \begin{pmatrix} \gamma & -\frac{u}{c}\gamma & 0 & 0 \\ -\frac{u}{c}\gamma & \gamma & 0 & 0 \\ 0 & 0 & 1 & 0 \\ 0 & 0 & 0 & 1 \end{pmatrix} \begin{pmatrix} ct \\ x \\ y \\ z \end{pmatrix}$$

$$\implies \boxed{x'^\mu = \Lambda^\mu{}_\nu x^\nu} \tag{8.39}$$

---

[4] When I say 4-vector, the "4" indicates dimension and when I say 2-tensor, the "2" indicates number of indices.

[5] The "1st language" of electrodynamics is vector calculus and this is like the 2nd language of electrodynamics. There is a 3rd language of electrodynamics which is the language of Differential Forms.

Electrodynamics Recast into a Manifestly Lorentz Covariant Form

Momentum 4-vector:

$$\begin{aligned} \frac{E'}{c} &= \gamma(u)\left(\frac{E}{c} - \frac{u}{c}p_x\right) \\ p'_x &= \gamma(u)\left(p_x - \frac{u}{c}\left(\frac{E}{c}\right)\right) \\ p'_y &= p_y \\ p'_z &= p_z \end{aligned} \implies \begin{pmatrix} E'/c \\ p'_x \\ p'_y \\ p'_z \end{pmatrix} = \begin{pmatrix} \gamma & -\frac{u}{c}\gamma & 0 & 0 \\ -\frac{u}{c}\gamma & \gamma & 0 & 0 \\ 0 & 0 & 1 & 0 \\ 0 & 0 & 0 & 1 \end{pmatrix} \begin{pmatrix} E/c \\ p_x \\ p_y \\ p_z \end{pmatrix}$$

$$\boxed{p'^{\mu} = \Lambda^{\mu}{}_{\nu} p^{\nu}}$$
(8.40)

Thus the Lorentz transformation of the field strength 2-tensor $F^{\mu\nu}$ is[6]

$$\boxed{F'^{\mu\nu} = \Lambda^{\mu}{}_{\sigma} \Lambda^{\nu}{}_{\rho} F^{\sigma\rho}} \tag{8.41}$$

| we write it into matrix multiplication format

$$F'^{\mu\nu} = \Lambda^{\mu}{}_{\sigma} F^{\sigma\rho} \Lambda^{\nu}{}_{\rho} \tag{8.42}$$

$$F'^{\mu\nu} = \Lambda^{\mu}{}_{\sigma} F^{\sigma\rho} (\Lambda^T){}_{\rho}{}^{\nu} \tag{8.43}$$

in matrix notation: $F' = \Lambda F \Lambda^T$ where Λ^T is the transpose of Λ (8.44)

It turns out that the correct arrangement of E_x, E_y, E_z, B_x, B_y and B_z in $F^{\mu\nu}$ so that $F' = \Lambda F \Lambda^T$ gives the same transformation results with the earlier "flying capacitor" derivation is

$$F^{\mu\nu} = \begin{pmatrix} 0 & E_x/c & E_y/c & E_z/c \\ -E_x/c & 0 & B_z & -B_y \\ -E_y/c & -B_z & 0 & B_x \\ -E_z/c & B_y & -B_x & 0 \end{pmatrix} \tag{8.45}$$

The field strength in S'-frame would simply be denoted as

$$F'^{\mu\nu} = \begin{pmatrix} 0 & E'_x/c & E'_y/c & E'_z/c \\ -E'_x/c & 0 & B'_z & -B'_y \\ -E'_y/c & -B'_z & 0 & B'_x \\ -E'_z/c & B'_y & -B'_x & 0 \end{pmatrix} \tag{8.46}$$

We quickly check 2 examples:

$$F'^{10} = \Lambda^1{}_{\sigma} \Lambda^0{}_{\rho} F^{\sigma\rho} \tag{8.47}$$

$$-\frac{E'_x}{c} = \Lambda^1{}_0 \Lambda^0{}_{\rho} F^{0\rho} + \Lambda^1{}_1 \Lambda^0{}_{\rho} F^{1\rho} \tag{8.48}$$

$$-\frac{E'_x}{c} = \Lambda^1{}_0 \Lambda^0{}_1 F^{01} + \Lambda^1{}_1 \Lambda^0{}_0 F^{10} \tag{8.49}$$

$$-\frac{E'_x}{c} = \left(-\frac{u}{c}\gamma\right)\left(-\frac{u}{c}\gamma\right)\left(\frac{E_x}{c}\right) + \gamma\gamma\left(-\frac{E_x}{c}\right) \tag{8.50}$$

$$-E'_x = -E_x\left(1 - \frac{u^2}{c^2}\right)\gamma^2 \tag{8.51}$$

$$E'_x = E_x \quad \text{(indeed)} \tag{8.52}$$

$$F'^{20} = \Lambda^2{}_{\sigma} \Lambda^0{}_{\rho} F^{\sigma\rho} \tag{8.53}$$

---

[6]At worst, you can take it as the recipe that for every index, you need one Lorentz matrix Λ to transform.

$$-\frac{E'_y}{c} = \Lambda^2{}_2 \Lambda^0{}_\rho F^{2\rho} \tag{8.54}$$

$$-\frac{E'_y}{c} = \Lambda^2{}_2 \left(\Lambda^0{}_0 F^{20} + \Lambda^0{}_1 F^{21} \right) \tag{8.55}$$

$$-\frac{E'_y}{c} = \gamma \left(-\frac{E_y}{c} \right) - \frac{u}{c}\gamma(-B_z) \tag{8.56}$$

$$E'_y = \gamma(E_y - uB_z) \quad \text{(indeed)} \tag{8.57}$$

You can also directly carry out matrix multiplication $F' = \Lambda F \Lambda^T$ to check every element.

Before we see Maxwell equations in covariant form, we first need to define a current density 4-vector:

Take an infinitesimal volume $\mathcal{V}$ which has charge q, the charge density is $\rho = \frac{q}{\mathcal{V}}$. Assume that $\mathcal{V}$ is moving along the x axis with speed u, the current density is $J_x = \rho u$. We define the rest charge density as $\rho_0 = \frac{q}{\mathcal{V}_0}$. Since only one dimension is contracted, $\mathcal{V} = \frac{1}{\gamma(u)}\mathcal{V}_0$ and so

$$\rho = \frac{q}{\mathcal{V}_0}\gamma(u) = \rho_0 \gamma(u) \tag{8.58}$$

$$J_x = \rho u = \rho_0 u \gamma(u) \tag{8.59}$$

We recall the (upper index) momentum 4-vector,

$$p^\mu = \left(\frac{E}{c}, p_x, p_y, p_z \right) = (mc\gamma(u_1), mu_1\gamma(u_1), 0, 0) \tag{8.60}$$

we are inspired to define the (upper index) current density 4-vector

$$\boxed{J^\mu = (c\rho, J_x, J_y, J_z) = (J^0, J^1, J^2, J^3)} = (\rho_0 c\gamma(u), \rho_0 u\gamma(u), 0, 0) \tag{8.61}$$

It must be noted that the charge continuity equation[7] is actually a Lorentz invariant equation.[8]

$$\frac{\partial \rho}{\partial t} + \vec{\nabla} \cdot \vec{J} = 0 \tag{8.62}$$

$$\frac{1}{c}\frac{\partial J^0}{\partial t} + \frac{\partial J_x}{\partial x} + \frac{\partial J_y}{\partial y} + \frac{\partial J_z}{\partial z} = 0 \tag{8.63}$$

$$\frac{\partial J^0}{\partial x^0} + \frac{\partial J^1}{\partial x^1} + \frac{\partial J^2}{\partial x^2} + \frac{\partial J^3}{\partial x^3} = 0 \tag{8.64}$$

$$\boxed{\partial_\mu J^\mu = 0} \quad \text{(charge continuity equation in Lorentz scalar form)} \tag{8.65}$$

Notice that the four Maxwell equations (see appendix 16) actually can be grouped into 2 types: (i) 2 equations that relates sources to the fields and (ii) 2 equations that are constraints on the fields. Thus it turns out that the Lorentz covariant Maxwell equations nicely group the 2 types of equations into 2 equations:[9]

$$\boxed{\partial_\nu F^{\mu\nu} = \mu_0 J^\mu \quad \text{and} \quad \partial^\sigma F^{\mu\nu} + \partial^\mu F^{\nu\sigma} + \partial^\nu F^{\sigma\mu} = 0} \tag{8.66}$$

---

[7] We call such an equation as "the 4-divergence of J is zero."

[8] It means it is a Lorentz scalar.

[9] Recall that the charge continuity equation is obtainable from Maxwell equations. In this covariant form, of course the same can be done. We take 4-divergence of $\partial_\nu F^{\mu\nu} = \mu_0 J^\mu$ to get $\partial_\mu \partial_\nu F^{\mu\nu} = \mu_0 \partial_\mu J^\mu$. The LHS is zero because $\partial_\mu \partial_\nu$ is symmetric and $F^{\mu\nu}$ is antisymmetric, so $\partial_\mu J^\mu = 0$.

Electrodynamics Recast into a Manifestly Lorentz Covariant Form

We check, for example, the component $\mu = 0$,

$$\partial_\nu F^{0\nu} = \mu_0 J^0 \tag{8.67}$$

$$\partial_0 F^{00} + \partial_1 F^{01} + \partial_2 F^{02} + \partial_3 F^{03} = \mu_0 c \rho \tag{8.68}$$

$$0 + \frac{\partial E_x/c}{\partial x} + \frac{\partial E_y/c}{\partial y} + \frac{\partial E_z/c}{\partial z} = \mu_0 c \rho \tag{8.69}$$

$$\frac{1}{c} \vec{\nabla} \cdot \vec{E} = \mu_0 c \rho \tag{8.70}$$

$$\text{recall that } \frac{1}{c^2} = \mu_0 \epsilon_0 \quad |$$

$$\vec{\nabla} \cdot \vec{E} = \frac{\rho}{\epsilon_0} \tag{8.71}$$

which is Gauss law. The other 3 Maxwell equations can be checked as exercise.

Next is the Lorentz force law. We already have a force 4-vector (or 4-force) called the Minkowski force K^μ.

$$K^\mu = \frac{dp^\mu}{d\tau} \tag{8.72}$$

In terms of the fields, or RHS of the law $\vec{F} = q(\vec{E} + (\vec{v} \times \vec{B}))$ we deduce the relativistic version by replacing $\vec{v} \longrightarrow u_\nu$ and $\vec{B} \longrightarrow F^{\mu\nu}$ giving the ansatz

$$\boxed{K^\mu = q F^{\mu\nu} u_\nu} \tag{8.73}$$

We check the spatial part,

$$K^1 = q F^{1\nu} u_\nu \tag{8.74}$$

$$K^1 = q F^{10} u_0 + q F^{11} u_1 + q F^{12} u_2 + q F^{13} u_3 \tag{8.75}$$

$$| \quad \text{note that } F^{11} = 0$$

$$| \quad \text{recall } u^\mu = \frac{p^\mu}{m} = \frac{1}{m}(mc\gamma(u), mu\gamma(u), 0, 0) \text{ where } u \text{ is in } x \text{ direction}$$

$$| \quad \text{assume } u \text{ has 3 components, so } u_\mu = \eta_{\mu\nu} u^\nu = (-c\gamma(u), u_x \gamma(u), u_y \gamma(u), u_z \gamma(u))$$

$$K^1 = q\gamma(u)\left(-c\left(-\frac{E_x}{c}\right) + u_y B_z + u_z(-B_y)\right) \tag{8.76}$$

$$K^1 = q\gamma(u)\left(\vec{E} + (\vec{u} \times \vec{B})\right)_x \tag{8.77}$$

then,

$$\vec{K} = \frac{d\vec{p}}{d\tau} = q\gamma(u)\left(\vec{E} + (\vec{u} \times \vec{B})\right) \tag{8.78}$$

$$| \quad \text{recall } dt = \gamma(u) d\tau$$

$$\frac{d\vec{p}}{dt} = q\left(\vec{E} + (\vec{u} \times \vec{B})\right) \tag{8.79}$$

where $\vec{p}$ is the relativistic momentum. The details of K^0 shall be worked out as an exercise.

Finally, we need to deal with the potentials: ϕ and $\vec{A}$. Turns out that they form the potential 4-vector (or 4-potential).

$$\boxed{A^\mu = \left(\frac{\phi}{c}, A_x, A_y, A_z\right)} \tag{8.80}$$

which means it has to transform as follows:

$$\frac{\phi'}{c} = \gamma(u)\left(\frac{\phi}{c} - \frac{u}{c}A_x\right) \tag{8.81}$$

$$A'_x = \gamma(u)\left(A_x - \frac{u}{c}\left(\frac{\phi}{c}\right)\right) \tag{8.82}$$

$$A'_y = A_y \tag{8.83}$$

$$A'_z = A_z \tag{8.84}$$

and we now check that indeed these transformations are compatible with the transformation properties of the fields. Recall $\vec{E} = -\vec{\nabla}\phi - \frac{\partial \vec{A}}{\partial t}$,

$$E_x = -\frac{\partial \phi}{\partial x} - \frac{\partial A_x}{\partial t} \tag{8.85}$$

$$E_x = \text{recall } \frac{\partial}{\partial x} = \partial^1 \text{ and } -\frac{1}{c}\frac{\partial}{\partial t} = \partial^0$$

$$E_x = -\partial^1 \phi + c\partial^0 A_x \tag{8.86}$$

$$\Longrightarrow E'_x = -\partial'^1 \phi' + c\partial'^0 A'_x \tag{8.87}$$

$$= -\Lambda^1{}_\nu \partial^\nu \left(\gamma(\phi - uA_x)\right) + c\Lambda^0{}_\mu \partial^\mu \left(\gamma\left(A_x - \frac{u}{c^2}\phi\right)\right) \tag{8.88}$$

$$E'_x = -\left(-\frac{u}{c}\gamma\partial^0 + \gamma\partial^1\right)\left(\gamma(\phi - uA_x)\right) + c\left(\gamma\partial^0 - \frac{u}{c}\gamma\partial^1\right)\left(\gamma\left(A_x - \frac{u}{c^2}\phi\right)\right) \tag{8.89}$$

| expand and 4 terms cancel pairwise

$$E'_x = -\gamma^2(\partial^0 A_x)\left(\frac{u^2}{c} - c\right) - \gamma^2(\partial^1 \phi)\left(1 - \frac{u^2}{c^2}\right) \tag{8.90}$$

| recall $\gamma^2 = \frac{1}{1 - \frac{u^2}{c^2}}$

$$E'_x = -\partial^1 \phi + c\partial^0 A_x \tag{8.91}$$

$$E'_x = E_x \text{ (indeed)} \tag{8.92}$$

We can check B_y, recall $\vec{B} = \vec{\nabla} \times \vec{A}$,

$$B_y = \frac{\partial A_x}{\partial z} - \frac{\partial A_z}{\partial x} = \partial^3 A_x - \partial^1 A_z \tag{8.93}$$

$$\Longrightarrow B'_y = \partial'^3 A'_x - \partial'^1 A'_z \tag{8.94}$$

$$B'_y = \partial^3 \left(\gamma\left(A_x - \frac{u}{c^2}\phi\right)\right) - \Lambda^1{}_\mu \partial^\mu A_z \tag{8.95}$$

$$B'_y = \gamma\left(\partial^3 A_x - \frac{u}{c^2}\partial^3 \phi\right) - \left(-\frac{u}{c}\gamma\partial^0 + \gamma\partial^1\right)A_z \tag{8.96}$$

| expand out and use $B_y = \partial^3 A_x - \partial^1 A_z$

| and $E_z = -\frac{\partial \phi}{\partial z} - \frac{\partial A_z}{\partial t} = -\partial^3 \phi + c\partial^0 A_z$

$$B'_y = \gamma\left(B_y + \frac{u}{c^2}E_z\right) \text{ (indeed)} \tag{8.97}$$

Note the very important pattern:

$$E_x = -\partial^1 \phi + c\partial^0 A_x \quad \text{and} \quad B_y = \partial^3 A_x - \partial^1 A_z$$
$$\Downarrow \qquad\qquad\qquad \Downarrow$$
$$cF^{01} = c\partial^0 A^1 - c\partial^1 A^0 \qquad F^{31} = \partial^3 A^1 - \partial^1 A^3$$
$$\Downarrow$$
$$F^{01} = \partial^0 A^1 - \partial^1 A^0$$

Thus, we get this general expression

$$\boxed{F^{\mu\nu} = \partial^\mu A^\nu - \partial^\nu A^\mu} \tag{8.98}$$

Recall the Lorenz gauge (section 2.2.2)

$$\vec{\nabla} \cdot \vec{A} = -\mu_0 \epsilon_0 \frac{\partial \phi}{\partial t} \tag{8.99}$$

$$\partial_1 A^1 + \partial_2 A^2 + \partial_3 A^3 = -\partial_0 A^0 \tag{8.100}$$

$$\boxed{\partial_\mu A^\mu = 0} \tag{8.101}$$

Thus the Lorenz gauge is Lorentz invariant. The general gauge transformation equations can also be put into a covariant form

$$\left. \begin{array}{l} \vec{A}' = \vec{A} + \vec{\nabla}\lambda \\ \phi' = \phi - \frac{\partial \lambda}{\partial t} \end{array} \right\} \quad \left. \begin{array}{l} A'^i = A^i + \partial^i \lambda \\ A'^0 = A^0 + \partial^0 \lambda \end{array} \right\} \quad \boxed{A'^\mu = A^\mu + \partial^\mu \lambda} \tag{8.102}$$

and the differential equations in the potential formulation (in Lorenz gauge) can also be written in a covariant form

$$\left. \begin{array}{l} \vec{\nabla}^2 \phi - \mu_0\epsilon_0 \frac{\partial^2 \phi}{\partial t^2} = -\frac{\rho}{\epsilon_0} \\ \vec{\nabla}^2 \vec{A} - \mu_0\epsilon_0 \frac{\partial^2 \vec{A}}{\partial t^2} = -\mu_0 \vec{J} \end{array} \right\} \; \left. \begin{array}{l} \Box(cA^0) = -\frac{J^0}{c\epsilon_0} \\ \Box A^i = -\mu_0 J^i \end{array} \right\} \; \left. \begin{array}{l} \Box A^0 = -\mu_0 J^0 \\ \Box A^i = -\mu_0 J^i \end{array} \right\} \boxed{\Box A^\mu = -\mu_0 J^\mu} \tag{8.103}$$

8.3.2 Conservation laws

We have already seen the rewriting of the charge continuity equation (8.65) into a Lorentz invariant (or scalar) form:

$$\text{Charge continuity equation: } \partial_\mu J^\mu = 0 \tag{8.104}$$

We have 2 other continuity equations for energy and momentum and from the definition of 4-momentum, we know that energy and momentum are closely related. Thus, we expect the 2 continuity equations to merge into one and has a similar form to the charge continuity equation, i.e. a 4-divergence form.

We recall from Part I sections 1.2 and 1.3:

$$\text{Energy continuity equation: } \vec{E} \cdot \vec{J} = -\frac{\partial u_{\text{EM}}}{\partial t} - \vec{\nabla} \cdot \vec{S} \tag{8.105}$$

$$\text{Momentum continuity equation: } \rho\vec{E} + \vec{J} \times \vec{B} = \vec{\nabla} \cdot \overset{\leftrightarrow}{T} - \frac{1}{c^2}\frac{\partial \vec{S}}{\partial t} \tag{8.106}$$

Since $\overset{\leftrightarrow}{T}$ is already a 3×3 matrix, we can go further and "promote" $\overset{\leftrightarrow}{T}$ into a 2-tensor $T^{\mu\nu}$. We have 4 objects to insert: u_{EM}, S_x, S_y and S_z. Since $\overset{\leftrightarrow}{T}$ is already symmetric, so $T^{\mu\nu}$ should also be symmetric[10], thus we try

$$\boxed{\text{Energy-momentum tensor: } T^{\mu\nu} = \begin{pmatrix} u_{\text{EM}} & S_x/c & S_y/c & S_z/c \\ S_x/c & & & \\ S_y/c & & -\overset{\leftrightarrow}{T} & \\ S_z/c & & & \end{pmatrix}} \tag{8.107}$$

---

[10]Einstein realised that $T^{\mu\nu}$ is the most general energy-momentum object to write and so he carried $T^{\mu\nu}$ to General Relativity and used it as the source of spacetime curvature (gravity). Also since the LHS of Einstein equation is a symmetric 2-tensor, it would match the number of components in $T^{\mu\nu}$.

We recall from Minkowski force: $K^1 = qF^{1\nu}u_\nu = q\gamma(u)\left(\vec{E} + (\vec{u}\times\vec{B})\right)_x$ so we try a similar expression $F^{\mu\nu}J_\nu$ and it should contain $\rho\vec{E} + \vec{J}\times\vec{B}$. So finally the ansatz for the combined continuity equation is

$$\boxed{\text{Energy-momentum continuity equation: } \partial_\mu T^{\mu\nu} = -F^{\nu\sigma}J_\sigma} \quad (8.108)$$

We check this ansatz. For component $\nu = 0$,

$$\partial_\mu T^{\mu 0} = -F^{0\sigma}J_\sigma \quad (8.109)$$
$$\partial_0 T^{00} + \partial_1 T^{10} + \partial_2 T^{20} + \partial_3 T^{30} = -\left(F^{01}J_1 + F^{02}J_2 + F^{03}J_3\right) \quad (8.110)$$
$$\frac{1}{c}\frac{\partial u_{\text{EM}}}{\partial t} + \frac{1}{c}\vec{\nabla}\cdot\vec{S} = -\frac{1}{c}\vec{E}\cdot\vec{J} \quad (8.111)$$

now for component $\nu = 1$,

$$\partial_\mu T^{\mu 1} = -F^{1\sigma}J_\sigma \quad (8.112)$$
$$\partial_0 T^{01} + \partial_1 T^{11} + \partial_2 T^{21} + \partial_3 T^{31} = -\left(F^{10}J_0 + F^{12}J_2 + F^{13}J_3\right) \quad (8.113)$$
$$\frac{1}{c^2}\frac{\partial S_x}{\partial t} + \left(\vec{\nabla}\cdot(-\overleftrightarrow{T})\right)_x = -\left(\left(-\frac{E_x}{c}\right)(-c\rho) + B_z J_y - B_y J_z\right) \quad (8.114)$$
$$\frac{1}{c^2}\frac{\partial S_x}{\partial t} + \left(\vec{\nabla}\cdot(-\overleftrightarrow{T})\right)_x = -\left(\rho E_x + (\vec{J}\times\vec{B})_x\right) \quad (8.115)$$

So indeed, $T^{\mu\nu}$ is the relativistic generalisation of $(-\overleftrightarrow{T})$. Lastly we must ensure that $T^{\mu\nu}$ is really a 2-tensor. We take $T_{ij} = \epsilon_0\left(E_i E_j - \frac{1}{2}\delta_{ij}\vec{E}^2\right) + \cdots$ and make a (really intelligent) guess that $T^{\mu\nu}$ should have a similar form.

$$\boxed{T^{\mu\nu} = -\frac{1}{\mu_0}\left(\eta^{\mu\sigma}F_{\sigma\alpha}F^{\alpha\nu} + \frac{1}{4}\eta^{\mu\nu}F_{\alpha\beta}F^{\alpha\beta}\right)} \quad (8.116)$$

We check

$$T^{00} = -\frac{1}{\mu_0}\left(\eta^{0\sigma}F_{\sigma\alpha}F^{\alpha 0} + \frac{1}{4}\eta^{00}F_{\alpha\beta}F^{\alpha\beta}\right) \quad (8.117)$$

$$\mid \text{ recall that } \eta^{00} = -1 \text{ and } F_{\alpha\beta}F^{\alpha\beta} = -\frac{2\vec{E}^2}{c^2} + 2\vec{B}^2$$

$$T^{00} = -\frac{1}{\mu_0}\left(-\frac{\vec{E}^2}{c^2} + \frac{1}{2}\frac{\vec{E}^2}{c^2} - \frac{1}{2}\vec{B}^2\right) \quad (8.118)$$

$$T^{00} = \frac{1}{2}\epsilon_0\vec{E}^2 + \frac{1}{2\mu_0}\vec{B}^2 \quad (8.119)$$

$$T^{00} = u_{\text{EM}} \quad (8.120)$$

In the exercise, you will check $F_{\alpha\beta}F^{\alpha\beta} = -\frac{2\vec{E}^2}{c^2} + 2\vec{B}^2$, $T^{0i} = \frac{S_i}{c}$ and $T^{ij} = -\epsilon_0\left(E_i E_j - \frac{1}{2}\delta_{ij}\vec{E}^2\right) + \cdots$. At the end of Part II, we will explicitly take 4-divergence of equation (8.116) to show its conservation law (eq (8.108)) so as to check for consistency.

Finally we just want to quickly mention the angular momentum 3-tensor,

$$\text{Angular momentum 3-tensor definition: } M^{\alpha\beta\gamma} = T^{\alpha\beta}x^\gamma - T^{\alpha\gamma}x^\beta \quad (8.121)$$
$$\text{Conservation law: } \partial_\alpha M^{\alpha\beta\gamma} = 0 \quad \text{(without external sources)} \quad (8.122)$$

8.3.3 Lorentz invariants in electrodynamics

We know that by forming "inner products", we can make Lorentz scalars which are invariant. They have the same numerical value in any inertial frame. Recall the following scalars we have seen earlier,

1. "inner product" of position 4-vector with itself gives the invariant interval
2. "inner product" of momentum 4-vector with itself gives the Einstein relation
3. "inner product" of derivative 4-vector with current density 4-vector gives the charge continuity equation

There are 2 invariants in electrodynamics. The first one is the obvious "inner product" of the field strength tensor with itself,

$$F^{\mu\nu}F_{\mu\nu} = 2\left(\vec{B}^2 - \frac{\vec{E}^2}{c^2}\right) \tag{8.123}$$

In the exercise, you will check some consequences of it and verify this invariant using the Lorentz transformations of $\vec{E}$ and $\vec{B}$.

The second invariant is much less obvious. It is deduced from the second Maxwell equation $\partial^\sigma F^{\mu\nu} + \partial^\mu F^{\nu\sigma} + \partial^\nu F^{\sigma\mu} = 0$ that we can define the so-called dual field strength tensor $\widetilde{F}^{\mu\nu}$ by

$$\widetilde{F}_{\mu\nu} = \frac{1}{2}\epsilon_{\mu\nu\alpha\beta}F^{\alpha\beta} \quad \text{with } \epsilon_{\mu\nu\alpha\beta} \text{ being the 4D Levi-Civita symbol} \tag{8.124}$$

so that the second Maxwell equation becomes $\boxed{\partial^\mu \widetilde{F}_{\mu\nu} = 0}$ which looks similar in form to the first Maxwell equation $\partial_\nu F^{\mu\nu} = \mu_0 J^\mu$. We check the 0-component,

$$\underbrace{\partial^\mu \widetilde{F}_{\mu 0}}_{=0} = \frac{1}{2}\epsilon_{\mu 0 \alpha\beta}\partial^\mu F^{\alpha\beta} \tag{8.125}$$

$$0 = \frac{1}{2}\Big(\epsilon_{1032}\partial^1 F^{32} + \epsilon_{1023}\partial^1 F^{23} + \epsilon_{3012}\partial^3 F^{12}$$
$$+ \epsilon_{3021}\partial^3 F^{21} + \epsilon_{2013}\partial^2 F^{13} + \epsilon_{2031}\partial^2 F^{31}\Big) \tag{8.126}$$

| the even permutations are ϵ_{3021}, ϵ_{1032} and ϵ_{2013}
| the odd permutations are ϵ_{3012}, ϵ_{1023} and ϵ_{2031}
| then use $F^{12} = -F^{21}$, $F^{23} = -F^{32}$ and $F^{13} = -F^{31}$

$$0 = -\left(\partial^1 F^{23} + \partial^3 F^{12} + \partial^2 F^{31}\right) \tag{8.127}$$
$$0 = \partial^1 F^{23} + \partial^3 F^{12} + \partial^2 F^{31} \tag{8.128}$$

Other components can be checked similarly. The second invariant is thus the "inner product" of $\widetilde{F}^{\mu\nu}$ with $F_{\mu\nu}$

$$\widetilde{F}^{\mu\nu}F_{\mu\nu} = \widetilde{F}_{\mu\nu}F^{\mu\nu} = \frac{1}{2}\epsilon_{\mu\nu\alpha\beta}F^{\alpha\beta}F^{\mu\nu} = \frac{4}{c}\vec{E}\cdot\vec{B} \tag{8.129}$$

which is to be checked in the exercise[11]. Also, in the exercise, you will also check some consequences of it and verify this invariant using the Lorentz transformations of $\vec{E}$ and $\vec{B}$.

---

[11] This term is called the "Axion" term. In the recent 15 years, condensed matter physics needs this term and uses it to describe 3D Topological Insulators.

8.4 Point charge revisited: Lienard–Wiechert potentials

Although all these rewriting of electrodynamics into Lorentz covariant form does not really contain any new physics (except maybe relativistic corrections), it does provide a whole new perspective and a new machinery to uncover more physics.

We shall rederive the Lienard–Wiechert potentials (i.e. retarded potentials for a moving point charge) in relativistic language.

First we need to rewrite the (retarded) Green's function into a Lorentz invariant form.

$$G^R(\vec{r},t;\vec{r}',t') = -\theta(t-t')\frac{1}{4\pi R}\delta\left(t-t'-\frac{R}{c}\right) \qquad (8.130)$$

| we need an identity which we will prove now
| start with $\delta\left((\bar{x}-\bar{x}')^2\right) = \delta\left(-(x^0-x'^0)^2 + |\vec{r}-\vec{r}'|^2\right)$
| $\qquad = \delta\left(R^2 - (x^0-x'^0)^2\right)$
| $\qquad = \delta\left((R-x^0+x'^0)(R+x^0-x'^0)\right)$
| $\qquad = \frac{1}{2R}\left[\delta(R-ct+ct') + \delta(R+ct-ct')\right]$
| $\qquad = \frac{1}{2Rc}\left[\delta\left(\frac{R}{c}-t+t'\right) + \delta\left(\frac{R}{c}+t-t'\right)\right]$
| and since delta function is even
| $\qquad = \frac{1}{2Rc}\left[\delta\left(t-t'-\frac{R}{c}\right) + \delta\left(t-t'+\frac{R}{c}\right)\right]$

$$= -\theta(t-t')\frac{c}{2\pi}\delta\left((\bar{x}-\bar{x}')^2\right) \qquad (8.131)$$

| note that the extra delta function is "negated" by the step function
| note that $\theta(t-t')$ has the same effect as $\theta(ct-ct') = \theta(x^0-x'^0)$

$$G^R(\bar{x},\bar{x}') = -\theta(x^0-x'^0)\frac{c}{2\pi}\delta\left((\bar{x}-\bar{x}')^2\right) \qquad (8.132)$$

So the step function must now be explicitly written because we got an extra delta function. The step function is invariant in the sense that if $t > t'$ in one inertial frame, then the "transformed t" > "transformed t'" in all other inertial frames.

Next is the 4-current density. We recall from section 3.1, the point charge density and the point charge current density

$$c\rho(\vec{r},t) = qc\delta^3(\vec{r}-\vec{r}_0(t)) \text{ and } \vec{J}(\vec{r},t) = q\vec{v}(t)\delta^3(\vec{r}-\vec{r}_0(t)) \qquad (8.133)$$

Note that we take all these coordinates and time to be measured in S-frame. In S-frame, we shall parameterise the charged particle's position 4-vector with the proper time τ,

$$(x_0)^\mu = (ct,\vec{r}_0(t)) = (ct(\tau),\vec{r}_0(t(\tau))) = (ct(\tau),\vec{r}_0(\tau)) \qquad (8.134)$$

$$\text{velocity 4-vector: } u^\mu = \frac{d(x_0)^\mu}{d\tau} = (\gamma c, \gamma\vec{v}) \text{ where } \vec{v} = \frac{d\vec{r}_0(t)}{dt} \qquad (8.135)$$

To make $J^\mu = (c\rho, \vec{J})$ manifestly covariant, we introduce an extra temporal delta function,

$$c\rho(\vec{r},t) = J^0 = qc\int d(t(\tau))\delta^3(\vec{r}-\vec{r}_0(\tau))\delta(t-t(\tau)) \qquad (8.136)$$

| recall that $dt = \gamma d\tau$ and put c into $\delta(t-t(\tau))$

Electrodynamics Recast into a Manifestly Lorentz Covariant Form 85

$$c\rho(\bar{x}) = J^0 = qc^2 \int d\tau \gamma \delta^3(\vec{r} - \vec{r}_0(\tau))\delta(ct - ct(\tau)) \tag{8.137}$$

$$|\quad \text{write } c\gamma = u^0$$

$$c\rho(\bar{x}) = J^0 = qc \int d\tau u^0 \delta^4(\bar{x} - \bar{x}_0) \tag{8.138}$$

$$\text{similarly, } \vec{J}(\vec{r},t) = q \int d(t(\tau)) \vec{v} \delta^3(\vec{r} - \vec{r}_0(\tau))\delta(t - t(\tau)) \tag{8.139}$$

$$\vec{J}(\vec{r},t) = qc \int d\tau \gamma \vec{v} \delta^4(\bar{x} - \bar{x}_0) \tag{8.140}$$

$$\text{so, } J^\mu(\bar{x}) = qc \int d\tau\, u^\mu \delta^4(\bar{x} - \bar{x}_0) \tag{8.141}$$

Now, the retarded potentials (recall $\Lambda^\mu = \left(\frac{\phi}{c}, \vec{A}\right)$),

$$\phi(\vec{r},t) = -\frac{1}{\epsilon_0} \int G^R(\vec{r},t;\vec{r}',t')\rho(\vec{r}',t')dV'dt' \tag{8.142}$$

$$|\quad \text{to write into } d^4\bar{x}, \text{ we convert } dt \text{ into unit of length}$$

$$\frac{\phi(\vec{r},t)}{c} = -\frac{1}{\epsilon_0 c} \int \left(-\theta(x^0 - x'^0)\frac{c}{2\pi}\delta((\bar{x} - \bar{x}')^2)\right) \frac{1}{c} J^0(\bar{x}') \frac{1}{c} d^4\bar{x}' \tag{8.143}$$

$$A^0(\bar{x}) = \frac{1}{2\pi\epsilon_0 c^2} \int \theta(x^0 - x'^0)\delta((\bar{x} - \bar{x}')^2) J^0(\bar{x}) d^4\bar{x}' \tag{8.144}$$

$$\text{then, } \vec{A}(\vec{r},t) = -\mu_0 \int G^R(\vec{r},t;\vec{r}',t')\vec{J}(\vec{r}',t')dV'dt' \tag{8.145}$$

$$\vec{A}(\vec{r},t) = -\mu_0 \int \left(-\theta(x^0 - x'^0)\frac{c}{2\pi}\delta((\bar{x} - \bar{x}')^2)\right) \vec{J}(\bar{x}') \frac{1}{c} d^4\bar{x}' \tag{8.146}$$

$$|\quad \text{recall that } \frac{1}{c^2} = \epsilon_0 \mu_0$$

$$\vec{A}(\vec{r},t) = \frac{1}{2\pi\epsilon_0 c^2} \int \theta(x^0 - x'^0)\delta((\bar{x} - \bar{x}')^2)\vec{J}(\bar{x}')d^4\bar{x}' \tag{8.147}$$

So the covariant form is[12]

$$A^\mu(\bar{x}) = \frac{1}{2\pi\epsilon_0 c^2} \int \theta(x^0 - x'^0)\delta((\bar{x} - \bar{x}')^2) J^\mu(\bar{x}')d^4\bar{x}' \tag{8.148}$$

$$|\quad \text{insert } J^\mu(\bar{x}') = qc \int d\tau u^\mu \delta^4(\bar{x}' - \bar{x}_0)$$

$$|\quad \text{then do the } d^4\bar{x}' \text{ integral using } \delta^4(\bar{x}' - \bar{x}_0)$$

$$A^\mu(\bar{x}) = \frac{q}{2\pi\epsilon_0 c} \int d\tau \theta(x^0 - (x_0)^0)\delta((\bar{x} - \bar{x}_0)^2) u^\mu \tag{8.149}$$

$$|\quad \text{the delta imposes the invariant interval } (\bar{x} - \bar{x}_0(\tau_r))^2 = 0$$

$$|\quad \text{which is called the light cone condition as zero invariant interval is}$$

$$\quad\quad \text{on the light cone!}$$

$$|\quad \text{the step function imposes the retardation condition } x^0 > (x_0)^0$$

$$|\quad \text{to evaluate further, we use the identity } \delta(f(\tau)) = \sum_i \frac{\delta(\tau - \tau_i)}{\left|\left(\frac{df}{d\tau}\right)_{\tau=\tau_i}\right|}$$

---

[12] Strictly speaking, we need to check that the volume element $d^4\bar{x}$ is a Lorentz scalar, i.e. the Jacobian = 1. This will be checked later when we discuss about the Lagrangian of the EM field.

$$\text{then } \frac{d}{d\tau}(\bar{x} - \bar{x}_0(\tau))^2 = -2(x - x_0(\tau))_\mu \frac{d(x_0)^\mu}{d\tau} = -2(x - x_0(\tau))_\mu u^\mu$$

$$\text{so } \delta(f(\tau)) = \frac{\delta(\tau - \tau_r)}{|-2(x - x_0(\tau))_\mu u^\mu|_{\tau = \tau_r}} \tag{8.150}$$

Note that $(\cdots)_\mu u^\mu$ is timelike and so is negative, thus

$$|-(\cdots)_\mu u^\mu| = -(\cdots)_\mu u^\mu$$

$$A^\mu(\bar{x}) = -\frac{q}{4\pi\epsilon_0 c} \frac{u^\mu}{(x - x_0(\tau))_\nu u^\nu}\bigg|_{\tau = \tau_r} \tag{8.151}$$

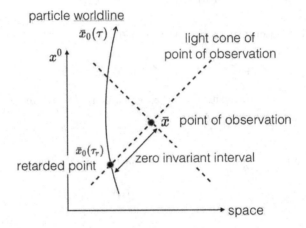

FIGURE 8.4
The 4D trajectory intersects the lightcone of the point of observation at 2 points. The retarded point is picked out by the step function and is the one of physical interest. Any "distance" (invariant interval) between 2 points on the light cone is zero.

8.5 Point charge revisited: Fields of a charged particle in constant velocity

We now revisit the calculation in section 3.3 where we calculated the electric and magnetic fields of a charge with constant velocity. In the context of special relativity, we simply need to have the charge in S'-frame and the point of observation to be in S-frame. S'-frame is moving with speed v along the x axis relative to S-frame.

We set the charged particle at the origin of S'-frame. The point of observation in S'-frame coordinates are (x', y', z'). As the charged particle is not moving with respect to S'-frame, there is only electrostatic electric field and no magnetic field, so B'_x, B'_y and $B'_z = 0$. The transformation of the fields become:

$$\begin{array}{lll} E'_x = E_x & E'_y = \gamma_v(E_y - vB_z) & E'_z = \gamma_v(E_z + vB_y) \\ 0 = B_x & 0 = \gamma_v\left(B_y + \frac{v}{c^2}E_z\right) & 0 = \gamma_v\left(B_z - \frac{v}{c^2}E_y\right) \end{array} \tag{8.152}$$

$$\text{so, } E'_y = \gamma_v \left(E_y - v \left(\frac{v}{c^2} E_y \right) \right) \tag{8.153}$$

$$E_y = \gamma_v E'_y \tag{8.154}$$

$$\text{and so, } E'_z = \gamma_v \left(E_z + v \left(-\frac{v}{c^2} E_z \right) \right) \tag{8.155}$$

$$E_z = \gamma_v E'_z \tag{8.156}$$

The Coulomb field as seen in S'-frame is $\vec{E}' = \frac{q}{4\pi\epsilon_0 r'^2} \hat{r}'$. Thus using spherical coordinates in S'-frame,

$$E'_x = \hat{x}' \cdot \vec{E}' = \sin\theta' \cos\phi' \hat{r}' \cdot \frac{q}{4\pi\epsilon_0 r'^2} \hat{r}' \tag{8.157}$$

$$E'_x = \frac{\sqrt{x'^2 + y'^2}}{r'} \frac{x'}{\sqrt{x'^2 + y'^2}} \frac{q}{4\pi\epsilon_0 r'^2} \tag{8.158}$$

$$E'_x = \frac{q}{4\pi\epsilon_0} \frac{x'}{(x'^2 + y'^2 + z'^2)^{3/2}} \tag{8.159}$$

$\quad\quad$ | insert $x' = \gamma_v(x - vt)$, $y' = y$ and $z' = z$

$$E_x = E'_x = \frac{q}{4\pi\epsilon_0} \frac{\gamma_v(x - vt)}{(\gamma_v^2(x - vt)^2 + y^2 + z^2)^{3/2}} \tag{8.160}$$

$$\text{similarly, } E_y = \gamma_v E'_y = \gamma_v \hat{y}' \cdot \vec{E}' \tag{8.161}$$

$$E_y = \frac{q}{4\pi\epsilon_0} \frac{\gamma_v y}{(\gamma_v^2(x - vt)^2 + y^2 + z^2)^{3/2}} \tag{8.162}$$

$$\text{and, } E_z = \gamma_v E'_z = \gamma_v \hat{z}' \cdot \vec{E} \tag{8.163}$$

$$E_z = \frac{q}{4\pi\epsilon_0} \frac{\gamma_v z}{(\gamma_v^2(x - vt)^2 + y^2 + z^2)^{3/2}} \tag{8.164}$$

So E_x obtained the gamma factor from coordinate transformation while E_y and E_z obtained the gamma factor from field transformation!

To compare with the corresponding example in section 3.3, we set the present position of the particle and the point of observation to be in the xy plane (so $z = 0$).

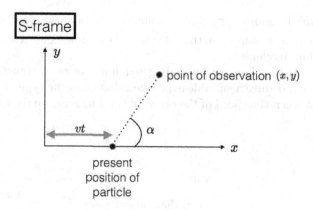

FIGURE 8.5
Setup where point of observation and present position of particle are in the xy plane of S-frame. We will shift the origin of S-frame to the present position of the particle.

We shift the origin to the present position of the particle and label the new x coordinate as $x_p = x - vt$. The position vector to the point of observation is $\vec{r}_p = x_p \hat{x} + y \hat{y}$ with $x_p = r_p \cos \alpha$ and $y = r_p \sin \alpha$.

$$\vec{E} = E_x \hat{x} + E_y \hat{y} \tag{8.165}$$

$$\vec{E} = \frac{q}{4\pi\epsilon_0} \gamma_v \left[\frac{x_p}{\left(\gamma_v^2 x_p^2 + y^2\right)^{3/2}} \hat{x} + \frac{y}{\left(\gamma_v^2 x_p^2 + y^2\right)^{3/2}} \hat{y} \right] \tag{8.166}$$

$$\vec{E} = \frac{q}{4\pi\epsilon_0} \frac{\gamma_v \vec{r}_p}{\left(\gamma_v^2 x_p^2 + y^2\right)^{3/2}} \tag{8.167}$$

$$\mid \text{ then, } \gamma_v^2 x_p^2 + y^2 = \frac{x_p^2}{1 - \frac{v^2}{c^2}} + y^2 = \frac{x_p^2 + y^2 - \frac{v^2}{c^2} y^2}{1 - \frac{v^2}{c^2}}$$

$$\mid \qquad = \gamma_v^2 \left(r_p^2 - \frac{v^2}{c^2} r_p^2 \sin^2 \alpha \right) = r_p^2 \gamma_v^2 \left(1 - \frac{v^2}{c^2} \sin^2 \alpha \right)$$

$$\vec{E} = \frac{q}{4\pi\epsilon_0} \frac{\gamma_v \vec{r}_p}{r_p^3 \gamma_v^3 \left(1 - \frac{v^2}{c^2} \sin^2 \alpha\right)^{3/2}} \tag{8.168}$$

$$\vec{E} = \frac{q}{4\pi\epsilon_0} \frac{1 - \frac{v^2}{c^2}}{\left(1 - \frac{v^2}{c^2} \sin^2 \alpha\right)^{3/2}} \frac{\hat{r}_p}{r_p^2} \tag{8.169}$$

which is exactly equation (3.60). Thus, the reason for the E field to be pointing from the present position instead of from the retarded position is because both E_x and E_y have the gamma factor and makes it a common factor and allows the expression to become $\gamma_v (x_p \hat{x} + y \hat{y}) \sim \gamma_v \vec{r}_p$. The gamma factors are obtained from different pieces of physics of coordinate transformation and field transformation.

8.6 Point charge revisited: Lienard's generalisation of Larmor's formula

In section 4.1, the Larmor's formula for point charge radiation is $P = \frac{\mu_0 q^2}{6\pi c} a^2$ and it is derived based on the assumption that it was at instantaneous rest or the particle was moving slowly (but accelerating).

It was generalised by Lienard by finding out how power P transforms under Lorentz transformations and deduce a suitable expression that suits the type of Lorentz object P is. Our check is that when the speed of the object v is set to zero, we recover Larmor's formula $P = \frac{\mu_0 q^2}{6\pi c} a^2$.

$$P = \frac{dE}{dt} = \frac{1}{\gamma} \frac{dE}{d\tau} \tag{8.170}$$

$$\mid \text{ recall that proper power is } K^0 = \frac{1}{c} \frac{dE}{d\tau}$$

$$P = \frac{1}{\gamma} c K^0 \tag{8.171}$$

Electrodynamics Recast into a Manifestly Lorentz Covariant Form

So if we guess P to be a Lorentz scalar

$$P = \frac{\mu_0 q^2}{6\pi c} a^2 \stackrel{\text{guess for } v \neq 0}{=} \frac{\mu_0 q^2}{6\pi c}(a^\nu a_\nu) \quad \text{where we define } a^\nu = \frac{du^\nu}{d\tau} \quad (8.172)$$

$$\text{then, } K^0 = \frac{\mu_0 q^2}{6\pi c^2}(a^\nu a_\nu)\gamma \quad (8.173)$$

| recall $u^0 = c\gamma$

$$K^0 = \frac{\mu_0 q^2}{6\pi c^3}(a^\nu a_\nu) u^0 \quad (8.174)$$

which properly makes K^0 a 0-component of a 4-vector. Thus power P indeed is a Lorentz scalar so now we simply need to work out $a_\nu a^\nu$ in terms of ordinary velocity and acceleration.

$$a^\nu = \frac{du^\nu}{d\tau} = \frac{dt}{d\tau}\frac{d}{dt}(c\gamma_v, \vec{v}\gamma_v) \quad (8.175)$$

$$a^\nu = \gamma_v\left(c\frac{d\gamma_v}{dt}, \frac{d\vec{v}}{dt}\gamma_v + \vec{v}\frac{d\gamma_v}{dt}\right) \quad (8.176)$$

| so $\dfrac{d\gamma_v}{dt} = \dfrac{d}{dt}\dfrac{1}{\sqrt{1-\frac{v^2}{c^2}}} = -\dfrac{1}{2\left(1-\frac{v^2}{c^2}\right)^{3/2}}\left(-\dfrac{2\vec{v}}{c^2}\cdot\dfrac{d\vec{v}}{dt}\right) = \gamma_v^3\dfrac{\vec{v}\cdot\vec{a}}{c^2}$

$$a^\nu = \left(\gamma_v^4\frac{\vec{v}\cdot\vec{a}}{c}, \gamma_v^2\vec{a} + \vec{v}\gamma_v^4\frac{\vec{v}\cdot\vec{a}}{c^2}\right) \quad (8.177)$$

$$a^\nu a_\nu = -(a^0)^2 + \vec{a}^2 \quad (8.178)$$

$$a^\nu a_\nu = -\gamma_v^8\frac{(\vec{v}\cdot\vec{a})^2}{c^2} + \gamma_v^4\left(\vec{a}+\vec{v}\gamma_v^2\frac{\vec{v}\cdot\vec{a}}{c^2}\right)\cdot\left(\vec{a}+\vec{v}\gamma_v^2\frac{\vec{v}\cdot\vec{a}}{c^2}\right) \quad (8.179)$$

$$a^\nu a_\nu = \gamma_v^4\left[-\gamma_v^4\frac{(\vec{v}\cdot\vec{a})^2}{c^2} + a^2 + 2\gamma_v^2\frac{(\vec{v}\cdot\vec{a})^2}{c^2} + v^2\gamma_v^4\frac{(\vec{v}\cdot\vec{a})^2}{c^4}\right] \quad (8.180)$$

$$a^\nu a_\nu = \gamma_v^4\left[a^2 + (\vec{v}\cdot\vec{a})^2\left(-\frac{1}{c^2\left(1-\frac{v^2}{c^2}\right)^2} + \frac{2}{c^2\left(1-\frac{v^2}{c^2}\right)} + \frac{v^2}{c^4\left(1-\frac{v^2}{c^2}\right)^2}\right)\right] \quad (8.181)$$

$$a^\nu a_\nu = \gamma_v^4\left[a^2 + (\vec{v}\cdot\vec{a})^2\left(-\frac{c^2}{(c^2-v^2)^2} + \frac{2}{c^2-v^2} + \frac{v^2}{(c^2-v^2)^2}\right)\right] \quad (8.182)$$

$$a^\nu a_\nu = \gamma_v^4\left[a^2 + \frac{(\vec{v}\cdot\vec{a})^2}{c^2-v^2}\right] \quad (8.183)$$

So the generalised power is

$$P = \frac{\mu_0 q^2}{6\pi c}\gamma_v^4\left[a^2 + \frac{(\vec{v}\cdot\vec{a})^2}{c^2-v^2}\right] \quad (8.184)$$

| we need to massage a bit to get to Lienard's form

$$P = \frac{\mu_0 q^2}{6\pi c}\gamma_v^6\left[a^2\left(1-\frac{v^2}{c^2}\right) + \frac{1}{c^2}(\vec{v}\cdot\vec{a})^2\right] \quad (8.185)$$

$$P = \frac{\mu_0 q^2}{6\pi c}\gamma_v^6\left[a^2 - \frac{1}{c^2}\left(v^2 a^2 - (\vec{v}\cdot\vec{a})^2\right)\right] \quad (8.186)$$

| write $\vec{v}\cdot\vec{a} = va\cos\theta$ so $v^2 a^2 - (\vec{v}\cdot\vec{a})^2 = v^2 a^2(1-\cos^2\theta) = v^2 a^2 \sin^2\theta = |\vec{v}\times\vec{a}|^2$

$$\boxed{P = \frac{\mu_0 q^2}{6\pi c}\gamma_v^6\left(a^2 - \frac{|\vec{v}\times\vec{a}|^2}{c^2}\right)} \quad (8.187)$$

This is Lienard's formula for the power of a radiating point charge. Obviously when we set $\vec{v} = 0$, we get Larmor's formula.

We shall now deal with the general expression of the power radiated per unit solid angle or $\frac{dP}{d\Omega}$. As we are now considering a moving charge, essentially the same geometrical effect in retarded potentials (section 3.1) also occurs here, i.e. the power radiated by the charge is not the power that is passing through the sphere of observation. They are related by the same geometrical factor $1 - \frac{\hat{R} \cdot \vec{v}}{c}$.

$$\vec{S}_{\text{passing through sphere}} = \left(1 - \frac{\hat{R} \cdot \vec{v}}{c}\right) \vec{S}_{\text{radiated by particle}} \quad (8.188)$$

For $\vec{v} = 0$, the Poynting vectors are the same. The expressions in Part 1 are actually $\vec{S}_{\text{radiated by particle}}$. Now we can continue the calculation for $\vec{v} \neq 0$ using equation (4.9) from Part I just before we set $\vec{v} = 0$,

$\vec{S}_{\text{passing through sphere}}$

$$= \left(1 - \frac{\hat{R} \cdot \vec{v}}{c}\right) \frac{1}{\mu_0 c} \left(\frac{q}{4\pi\epsilon_0} \frac{R}{(Rc - \vec{R} \cdot \vec{v})^3}\right)^2 \left[\vec{R} \times ((\hat{R}c - \vec{v}) \times \vec{a})\right] \cdot \left[\vec{R} \times ((\hat{R}c - \vec{v}) \times \vec{a})\right] \hat{R}$$

| write $1 - \frac{\hat{R} \cdot \vec{v}}{c} = 1 - \frac{\vec{R} \cdot \vec{v}}{Rc} = \frac{Rc - \vec{R} \cdot \vec{v}}{Rc}$

| also write $\left[\vec{R} \times ((\hat{R}c - \vec{v}) \times \vec{a})\right] \cdot \left[\vec{R} \times ((\hat{R}c - \vec{v}) \times \vec{a})\right] = R^2 \left|\hat{R} \times ((\hat{R}c - \vec{v}) \times \vec{a})\right|^2$

$$= \frac{1}{\mu_0 c^2} \frac{q^2}{16\pi^2 \epsilon_0^2} \frac{R^3 \left|\hat{R} \times ((\hat{R}c - \vec{v}) \times \vec{a})\right|^2}{(Rc - \vec{R} \cdot \vec{v})^5} \hat{R} \quad (8.189)$$

$$\frac{dP}{d\Omega} = \vec{S}_{\text{passing through sphere}} \cdot \hat{R} R^2 \quad (8.190)$$

$$\frac{dP}{d\Omega} = \frac{q^2}{16\pi^2 \epsilon_0} \frac{R^5 \left|\hat{R} \times ((\hat{R}c - \vec{v}) \times \vec{a})\right|^2}{(Rc - \vec{R} \cdot \vec{v})^5} \quad (8.191)$$

$$\frac{dP}{d\Omega} = \frac{q^2}{16\pi^2 \epsilon_0} \frac{\left|\hat{R} \times ((\hat{R}c - \vec{v}) \times \vec{a})\right|^2}{(c - \hat{R} \cdot \vec{v})^5} \quad (8.192)$$

We shall now consider 2 cases: (i) where $\vec{v}$ and $\vec{a}$ are (instantaneously) colinear and (ii) where $\vec{v}$ and $\vec{a}$ are (instantaneously) perpendicular to each other.[13]

---

[13] The logic for this entire section may seem messy for a reason. The natural logic should be:

1. include geometrical factor
2. calculate $\frac{dP}{d\Omega}$
3. integrate over sphere to get total power P
4. work out 2 special cases of $\vec{a} \parallel \vec{v}$ and $\vec{a} \perp \vec{v}$

The integration over sphere to get P is too difficult to carry out so I changed the logic to:

1. deduce total power P is a Lorentz scalar
2. generalise acceleration to a 4-vector so that P becomes a Lorentz scalar
3. then include geometrical factor to get $\frac{dP}{d\Omega}$
4. work out 2 special cases of $\vec{a} \parallel \vec{v}$ and $\vec{a} \perp \vec{v}$

Case of $\vec{a} \parallel \vec{v}$: [14] When $\vec{a}$ and $\vec{v}$ are parallel, $(\hat{R}c - \vec{v}) \times \vec{a} = c\hat{R} \times \vec{a}$ so,

$$\frac{dP}{d\Omega} = \frac{q^2}{16\pi^2\epsilon_0} \frac{\left|\hat{R} \times (c\hat{R} \times \vec{a})\right|^2}{(c - \hat{R}\cdot\vec{v})^5} \tag{8.193}$$

| use vector identity $\vec{A} \times (\vec{B} \times \vec{C}) = \vec{B}(\vec{A}\cdot\vec{C}) - \vec{C}(\vec{A}\cdot\vec{B})$
| so that $\hat{R} \times (\hat{R} \times \vec{a}) = \hat{R}(\hat{R}\cdot\vec{a}) - \vec{a}(\hat{R}\cdot\hat{R})$

$$\frac{dP}{d\Omega} = \frac{q^2 c^2}{16\pi^2\epsilon_0} \frac{\left|\hat{R}(\hat{R}\cdot\vec{a}) - \vec{a}\right|^2}{(c - \hat{R}\cdot\vec{v})^5} \tag{8.194}$$

| and $\left|\hat{R}(\hat{R}\cdot\vec{a}) - \vec{a}\right|^2 = (\hat{R}\cdot\vec{a})^2 + a^2 - 2(\hat{R}\cdot\vec{a})^2 = a^2 - (\hat{R}\cdot\vec{a})^2$
| let $\vec{v}$ point along the z axis, $\hat{R}\cdot\vec{v} = v\cos\theta$
| and $\hat{R}\cdot\vec{a} = \pm a\cos\theta$, negative if decelerating

$$\frac{dP}{d\Omega} = \frac{q^2 c^2}{16\pi^2\epsilon_0} \frac{a^2 \sin^2\theta}{c^5 \left(1 - \frac{v}{c}\cos\theta\right)^5} \tag{8.195}$$

| then use $c^2\epsilon_0 = \dfrac{1}{\mu_0}$

$$\frac{dP}{d\Omega} = \frac{\mu_0 q^2 a^2}{16\pi^2 c} \frac{\sin^2\theta}{\left(1 - \frac{v}{c}\cos\theta\right)^5} \tag{8.196}$$

We can integrate this to get the total power but we will apply Lienard's formula instead. Simply write $\vec{v} \times \vec{a} = 0$ and get

$$P = \frac{\mu_0 q^2 a^2}{6\pi c} \gamma_v^6 \tag{8.197}$$

Note that the expressions for $\frac{dP}{d\Omega}$ and P are applicable for both $\vec{a}$ parallel to $\vec{v}$ (accelerating) and $\vec{a}$ anti-parallel to $\vec{v}$ (decelerating). The lobes (of power) are stretched towards the direction of $\vec{v}$ (here, the z axis) although there is no radiation exactly in the direction of $\vec{v}$. A typical example is bremsstrahlung where an electron hits a metal and decelerates greatly with the emission of radiation.

Case of $\vec{a} \perp \vec{v}$: [15] We shall set $\vec{v}$ along the z axis, $\vec{a}$ along the x axis so

$$\vec{v} = v\hat{z}, \vec{a} = a\hat{x} \quad \text{and} \quad \hat{R} = \sin\theta\cos\phi\hat{x} + \sin\theta\sin\phi\hat{y} + \cos\theta\hat{z} \tag{8.198}$$

and we shall calculate $\frac{dP}{d\Omega}$ first.

$$\frac{dP}{d\Omega} = \frac{q^2}{16\pi^2\epsilon_0} \frac{\left|\hat{R} \times ((c\sin\theta\cos\phi\hat{x} + c\sin\theta\sin\phi\hat{y} + (c\cos\theta - v)\hat{z}) \times a\hat{x})\right|^2}{(c - v\cos\theta)^5} \tag{8.199}$$

---

[14] You may take $\vec{a} \parallel \vec{v}$ at an instant where time is t_r.
[15] You may take $\vec{a} \perp \vec{v}$ at an instant where time is t_r.

$$\frac{dP}{d\Omega} = \frac{q^2 a^2}{16\pi^2 \epsilon_0} \frac{\left|\hat{R} \times (-c\sin\theta\sin\phi\hat{z} + (c\cos\theta - v)\hat{y})\right|^2}{c^5 \left(1 - \frac{v}{c}\cos\theta\right)^5} \tag{8.200}$$

| divide c^2 in both numerator and denominator

| the cross product is $\begin{vmatrix} \hat{x} & \hat{y} & \hat{z} \\ \sin\theta\cos\phi & \sin\theta\sin\phi & \cos\theta \\ 0 & \cos\theta - \frac{v}{c} & -\sin\theta\sin\phi \end{vmatrix}$

| which $= -\left(\sin^2\theta\sin^2\phi + \cos\theta\left(\cos\theta - \frac{v}{c}\right)\right)\hat{x}$

| $\qquad - \sin^2\theta\sin\phi\cos\phi\hat{y} + \sin\theta\cos\phi\left(\cos\theta - \frac{v}{c}\right)\hat{z}$

$$\frac{dP}{d\Omega} = \frac{q^2 a^2}{16\pi^2\epsilon_0 c^3} \frac{1}{\left(1 - \frac{v}{c}\cos\theta\right)^5}\left(\left(\sin^2\theta\sin^2\phi + \cos\theta\left(\cos\theta - \frac{v}{c}\right)\right)^2 \right.$$
$$\left. + \sin^4\theta\sin^2\phi\cos^2\phi + \sin^2\theta\cos^2\phi\left(\cos\theta - \frac{v}{c}\right)^2\right)$$

| terms without $\frac{v}{c} = \sin^4\theta\sin^4\phi + \cos^4\theta + 2\sin^2\theta\sin^2\phi\cos^2\theta$
| $\qquad + \sin^4\theta\sin^2\phi\cos^2\phi + \sin^2\theta\cos^2\phi\cos^2\theta$
| |then 1st + 4th $= \sin^4\theta\sin^2\phi(\sin^2\phi + \cos^2\phi) = \sin^4\theta\sin^2\phi$
| |then 3rd + 5th $= \sin^2\theta\cos^2\theta(2\sin^2\phi + \cos^2\phi) = \sin^2\theta\cos^2\theta(\sin^2\phi + 1)$
| then terms without $\frac{v}{c} = \sin^4\theta\sin^2\phi + \cos^4\theta + \sin^2\theta\cos^2\theta\sin^2\phi + \sin^2\theta\cos^2\theta$
| |then 1st + 3rd $= \sin^2\theta\sin^2\phi(\sin^2\theta + \cos^2\theta) = \sin^2\theta\sin^2\phi$
| |then 2nd + 4th $= \cos^2\theta(\cos^2\theta + \sin^2\theta) = \cos^2\theta$
| then terms without $\frac{v}{c} = \sin^2\theta\sin^2\phi + \cos^2\theta$
| $\qquad = \sin^2\theta(1 - \cos^2\phi) + \cos^2\theta = 1 - \sin^2\theta\cos^2\phi$
| terms with $\frac{v}{c} = -2\frac{v}{c}\cos^3\theta - 2\frac{v}{c}\sin^2\theta\sin^2\phi\cos\theta - 2\frac{v}{c}\sin^2\theta\cos\theta\cos^2\phi$
| $\qquad = -2\frac{v}{c}(\cos^3\theta + \sin^2\theta\cos\theta(\sin^2\phi + \cos^2\phi))$
| $\qquad = -2\frac{v}{c}\cos\theta(\cos^2\theta + \sin^2\theta) = -2\frac{v}{c}\cos\theta$
| terms with $\frac{v^2}{c^2} = \frac{v^2}{c^2}\cos^2\theta + \frac{v^2}{c^2}\sin^2\theta\cos^2\phi$
| then write $\epsilon_0 c^2 = \frac{1}{\mu_0}$

$$\frac{dP}{d\Omega} = \frac{\mu_0 q^2 a^2}{16\pi^2 c} \frac{1 - \sin^2\theta\cos^2\phi - 2\frac{v}{c}\cos\theta + \frac{v^2}{c^2}\cos^2\theta + \frac{v^2}{c^2}\sin^2\theta\cos^2\phi}{\left(1 - \frac{v}{c}\cos\theta\right)^5} \tag{8.201}$$

| the 3 terms, $1 - 2\frac{v}{c}\cos\theta + \frac{v^2}{c^2}\cos^2\theta = \left(1 - \frac{v}{c}\cos\theta\right)^2$

$$\frac{dP}{d\Omega} = \frac{\mu_0 q^2 a^2}{16\pi^2 c} \frac{\left(1 - \frac{v}{c}\cos\theta\right)^2 - \left(1 - \frac{v^2}{c^2}\right)\sin^2\theta\cos^2\phi}{\left(1 - \frac{v}{c}\cos\theta\right)^5} \tag{8.202}$$

We can integrate this to get the total power but we apply Lienard's formula instead.

$$P = \frac{\mu_0 q^2}{6\pi c} \gamma_v^6 \left(a^2 - \frac{|\vec{v} \times \vec{a}|^2}{c^2} \right) \tag{8.203}$$

where $\vec{v} \times \vec{a} = va\hat{y}$

$$P = \frac{\mu_0 q^2 a^2}{6\pi c} \gamma_v^6 \left(1 - \frac{v^2}{c^2} \right) \tag{8.204}$$

$$P = \frac{\mu_0 q^2 a^2}{6\pi c} \gamma_v^4 \tag{8.205}$$

We see that when we set $\theta = 0$, $\frac{dP}{d\Omega}$ is sharply peaked in the direction of $\vec{v}$ for $\frac{v}{c} \to 1$. A typical example is circular motion where the electron emits (so-called) synchrotron radiation tangentially to its circular trajectory.

9

Exercises for Part II

9.1 Poincare group

We have seen in Part II that Lorentz transformations leave the invariant interval unchanged. Lorentz transformation is a continuous transformation because the relative velocity between the 2 inertial frames can take a continuous range of numbers. Now find out and show that the invariant interval is also unchanged under 2 other continuous transformations and 2 discrete transformations.

9.2 Minkowski diagram exercises

1. **Invariant hyperbola:**

    (a) In S-frame (of 1 time and 1 space), draw the locus of points where the invariant interval between 2 events $= -1$ by taking event 1 to be at the origin of S-frame and event 2 as all events with invariant interval -1 with respect to event 1. Any asymptotic lines must be drawn as well.

    (b) Draw the ct' axis of an S'-frame that is moving at $0.7c$ with respect to S-frame. This ct' axis will intersect the locus once and we shall call that intersection point to be event $2'$. State the significance of event 1 and event $2'$ with respect to S'-frame. State the significance (with respect to S'-frame) of the tangent (to the locus) at event $2'$.

2. **Relative simultaneity:**

    (a) If 2 events are separated by a spacelike interval, show that
        i. There exists a Lorentz frame in which they are simultaneous, and
        ii. in no Lorentz frame do they occur at the same (space) point.

    (b) If 2 events are separated by a timelike interval, show that
        i. There exists a frame in which they happen at the same (space) point, and
        ii. in no Lorentz frame are they simultaneous.

3. **Pole and Barn Paradox:** Take a barn to be stationary with respect to S-frame. The length of the barn is seen to be 1 m in S-frame. The left door (LD) of the barn is at the origin of the S-frame. Take S'-frame to be moving to the right with speed u along the length of the barn. A pole is seen to be stationary in S'-frame and seen to have a length of 1 m in S'-frame. The right end (RE) of the pole is at the origin of S'-frame. At $t = 0$, the origins of the 2 frames coincide, i.e. the event RE-LD occurs.

Exercises for Part II 95

The paradox is that, in S-frame, the pole is seen to be contracted and it fits entirely into the barn whereas in S'-frame, the barn is seen to be contracted and the pole does not fit entirely.

To resolve the paradox, draw a Minkowski diagram to determine the sequence of 2 events: (i) RE-RD (RD = right door) and (ii) LE-LD (LE = left end), in S-frame and S'-frame.

9.3 Photon

1. Suppose a photon has position 4-vector $x^\mu = (ct, x) = (3 \times 10^8, -3 \times 10^8)$ at $t = 1$ s, write down its position 4-vector at $t = 3$ s. Assume the photon is moving in the positive x direction.

2. The photon has zero rest mass so we need to be careful when writing down the 4-momentum of the photon. Take a photon to have energy E, write down the 4-momentum p^μ of the photon in terms of E. Assume the photon is moving in the positive x direction. (Hint: Assume Einstein relation still holds for the photon.)

9.4 Checking Lorentz transformations

Start with $F'^{\mu\nu} = \Lambda^\mu{}_\sigma \Lambda^\nu{}_\rho F^{\sigma\rho}$ and check

$$E'_z = \gamma(E_z + uB_y) \quad \text{and} \quad B'_y = \gamma\left(B_y + \frac{u}{c^2}E_z\right) \tag{9.1}$$

9.5 Maxwell's equations

1. Obtain Ampere–Maxwell law from $\partial_\nu F^{\mu\nu} = \mu_0 J^\mu$.
2. Obtain the remaining 2 Maxwell's equations from $\partial^\sigma F^{\mu\nu} + \partial^\mu F^{\nu\sigma} + \partial^\nu F^{\sigma\mu} = 0$.

9.6 Meaning of K^0

For $K^\mu = qF^{\mu\nu}u_\nu$, work out the 0-component in terms of time t and ordinary velocity $\vec{u}$ and explain its physical meaning.

9.7 Gauge invariance

Recall that gauge transformation takes the form $A'^\mu = A^\mu + \partial^\mu \lambda$. Check that $F^{\mu\nu} = \partial^\mu A^\nu - \partial^\nu A^\mu$ is gauge invariant.

9.8 Lorentz invariants

1. By using the Lorentz transformations of $\vec{E}$ and $\vec{B}$, show that $\vec{B}^2 - \frac{\vec{E}^2}{c^2}$ and $\vec{E} \cdot \vec{B}$ are indeed Lorentz invariants.

2. Suppose $\vec{E} \cdot \vec{B} = 0$, show that there is a Lorentz transformation which makes $\vec{E} = 0$ if $\vec{B}^2 - \frac{\vec{E}^2}{c^2} > 0$ and one that makes $\vec{B} = 0$ if $\vec{B}^2 - \frac{\vec{E}^2}{c^2} < 0$. What if $\vec{B}^2 - \frac{\vec{E}^2}{c^2} = 0$ in addition to $\vec{E} \cdot \vec{B} = 0$?

3. Take the $F^{\mu\nu}$ matrix and write down the characteristic equation (but do not solve it!). By observing the characteristic equation, give an argument why there are only 2 Lorentz invariants in electrodynamics. (Hint: Normally, you write the characteristic equation in Euclidean space. Here, we need to write the characteristic equation in Minkowski space.)

9.9 Lienard–Wiechert potentials in covariant form

Starting with the potentials in Part I,

$$\phi(\vec{r},t) = \frac{qc}{4\pi\epsilon_0} \frac{1}{R(t')c - \vec{R}(t') \cdot \vec{v}(t')}\bigg|_{t'=t_r} \quad \text{and} \quad \vec{A}(\vec{r},t) = \frac{q\mu_0}{4\pi} \frac{c\vec{v}(t')}{R(t')c - \vec{R}(t') \cdot \vec{v}(t')}\bigg|_{t'=t_r} \quad (9.2)$$

Show that it can be written into the covariant form (do not do it the other way round),

$$A^\mu(\bar{x}) = -\frac{q}{4\pi\epsilon_0 c} \frac{u^\mu}{(x - x_0(\tau))_\nu u^\nu}\bigg|_{\tau=\tau_r} \quad (9.3)$$

9.10 Radiation for acceleration parallel to velocity

Find the angle $\theta_{\max}$ (with respect to the velocity) at which the maximum radiation is emitted for the case of a charged particle with acceleration $\parallel$ to velocity. Show that for $v \approx c$, $\theta_{max} \approx \sqrt{\frac{1}{2}\left(1 - \frac{v}{c}\right)}$.

Part III

Classical Relativistic $U(1)$ Gauge Theory

10

Lagrangian Description of a Classical Relativistic $U(1)$ Gauge Theory

In year-1 undergraduate education, we learn physics from bottom-up, starting from Newton's laws and then in year-2, we learn analytical mechanics which is the top level of theory: Action Principle, Lagrangian and Hamiltonian mechanics and Noether's theorem. Classical mechanics is then done.

For electrodynamics, the same will take place now. We will develop the Lagrangian and Hamiltonian versions of electrodynamics and thus end off electrodynamics at the highest level of theory. This is also called "Classical Field Theory".

10.1 Review of Lagrangian mechanics and Noether's theorem

For a classical system described by n-generalised coordinates: $q_1(t), \ldots, q_n(t)$ and their corresponding velocities $\dot{q}_1(t), \ldots, \dot{q}_n(t)$, the Lagrangian is defined by

$$L(q_1(t), \ldots, q_n(t); \dot{q}_1(t), \ldots, \dot{q}_n(t); t) = T - V \qquad (10.1)$$

where T is the KE and V is the PE. Note that in field theory (where we deal with infinite degrees of freedom), the form $L = T - V$ may not really be valid as we will see later.

Hamilton's principle states that the trajectory in configuration space that extremises (maximum or minimum) the action $S = \int_{t_i}^{t_f} L\,dt$ is the actual (or physical) trajectory of the system.[1]

$$\text{Extremum means: } \delta S = 0 \qquad (10.2)$$

$$\implies \delta \int_{t_i}^{t_f} L(q_1(t), \ldots, q_n(t); \dot{q}_1(t), \ldots, \dot{q}_n(t); t)\,dt = 0 \qquad (10.3)$$

The requirement of extremum translates into n Euler-Lagrange equations of motion.

$$\frac{\partial L}{\partial q_i} - \frac{d}{dt}\left(\frac{\partial L}{\partial \dot{q}_i}\right) = 0 \qquad (10.4)$$

The next thing we want to mention is Noether's theorem which is the formal way of deriving conservation laws in a theory. Noether's theorem states: for each symmetry of the theory (i.e. transformations that does not change the Lagrangian), there is a conserved quantity associated to it.

---

[1] We assume fixed end points at t_i and t_f.

Suppose the Lagrangian is invariant under a change of coordinates with a small parameter ϵ,

$$\text{1st order change: } q'_i = q_i + \epsilon K_i(q) \tag{10.5}$$

where $K_i(q)$ may be a function of all the q_is, which we collectively denote as q.

We take the Lagrangian to be invariant to 1st order of the transformation,

$$L(q', \dot{q}') = L(q + \epsilon K, \dot{q} + \epsilon \dot{K}) \tag{10.6}$$

$$L(q', \dot{q}') \approx L(q, \dot{q}) + \sum_i \left(\frac{\partial L(q, \dot{q})}{\partial q_i} \epsilon K_i + \frac{\partial L(q, \dot{q})}{\partial \dot{q}_i} \epsilon \dot{K}_i \right) \tag{10.7}$$

$$\mid \text{ but } 0 = \delta L(q, \dot{q}) = L(q', \dot{q}') - L(q, \dot{q}) \tag{10.8}$$

$$0 = \sum_i \left(\frac{\partial L(q, \dot{q})}{\partial q_i} \epsilon K_i + \frac{\partial L(q, \dot{q})}{\partial \dot{q}_i} \epsilon \dot{K}_i \right) \tag{10.9}$$

$$\mid \text{ from Euler Lagrange, } \frac{\partial L}{\partial q_i} = \frac{d}{dt}\left(\frac{\partial L}{\partial \dot{q}_i}\right)$$

$$0 = \sum_i \left(\frac{d}{dt}\left(\frac{\partial L}{\partial \dot{q}_i}\right) K_i + \frac{\partial L}{\partial \dot{q}_i} \frac{dK_i}{dt} \right) \tag{10.10}$$

$$0 = \frac{d}{dt}\left(\sum_i \frac{\partial L}{\partial \dot{q}_i} K_i \right) \tag{10.11}$$

so the conserved quantity[2] is $\sum_i \frac{\partial L}{\partial \dot{q}_i} K_i$.

10.2 Relativistic Lagrangian of a charged particle interacting with external EM field

We start with a free (i.e. non-interacting) relativistic particle for simplicity. However $L = T = \frac{1}{2} m \frac{d\vec{x}}{dt} \cdot \frac{d\vec{x}}{dt} = \frac{1}{2} m v^2$ is obviously not compatible with special relativity. We have to make an intelligent guess of the free relativistic particle Lagrangian. To do that, we need to lay out the features that it should have.

- Following axiom 1 of Special Relativity, the action S should be a Lorentz scalar so that the action gives us the same physics in every inertial frame. Then we can write $S = \int L dt \overset{\gamma d\tau = dt}{\Longrightarrow} \int \gamma L d\tau$ and since proper time τ is invariant, γL should be a Lorentz scalar.

- In the non-relativistic limit $v \ll c$, we should have the familiar form: $L = \frac{1}{2} m v^2$.

- The Hamiltonian can be obtained by a Legendre transform of the Lagrangian L. From Einstein relation, we know the energy of a relativistic particle is $E^2 = \vec{p}^2 c^2 + m^2 c^4$, so the Hamiltonian is expected to be $H = \sqrt{\vec{p}^2 c^2 + m^2 c^4}$. Note that this expression is exact.

---

[2]The case of cyclic coordinates means its conjugate momenta is conserved, is a special case. Suppose the Lagrangian does not depend on q_3, so $q'_3 = q_3 + \epsilon$ does not change the Lagrangian. Thus $K_3 = 1$

$$\frac{d}{dt}\left(\frac{\partial L}{\partial \dot{q}_3} K_3\right) = 0 \implies \frac{dp_3}{dt} = 0 \tag{10.12}$$

We take $\gamma L = \alpha$ where α is some scalar and determine α by taking $v \ll c$,

$$L = \frac{\alpha}{\gamma} = \alpha\sqrt{1 - \frac{v^2}{c^2}} \approx \alpha\left(1 - \frac{v^2}{2c^2}\right) = \alpha - \frac{1}{2}\frac{\alpha}{c^2}v^2 \tag{10.13}$$

We require $L = \frac{1}{2}mv^2$, so $\alpha = -mc^2$.[3] Thus the relativistic Lagrangian is

$$\boxed{L = -mc^2\sqrt{1 - \frac{v^2}{c^2}}} \tag{10.14}$$

The equations of motion are obtained from the Euler-Lagrange equations:

$$\boxed{\frac{\partial L}{\partial \vec{x}} - \frac{d}{dt}\left(\frac{\partial L}{\partial \vec{v}}\right) = 0} \tag{10.15}$$

$$\text{so,}\quad \frac{d}{dt}\left(\frac{\partial L}{\partial \vec{v}}\right) = 0 \tag{10.16}$$

$$-mc^2 \frac{d}{dt}\left(\frac{\partial}{\partial \vec{v}}\sqrt{1 - \frac{v^2}{c^2}}\right) = 0 \tag{10.17}$$

$$-mc^2 \frac{d}{dt}\left(\frac{-\frac{2\vec{v}}{c^2}}{2\sqrt{1 - \frac{v^2}{c^2}}}\right) = 0 \tag{10.18}$$

$$\frac{d}{dt}(m\vec{v}\gamma) = 0 \tag{10.19}$$

So the conjugate momentum is $\vec{p} = \frac{\partial L}{\partial \vec{v}} = m\vec{v}\gamma$ which is the relativistic momentum. The Hamiltonian is derived from the Legendre transform,

$$H = \dot{\vec{x}} \cdot \vec{p} - L \tag{10.20}$$

$$H = \vec{v} \cdot (m\vec{v}\gamma) + mc^2\sqrt{1 - \frac{v^2}{c^2}} \tag{10.21}$$

$$\bigg|\ \text{from}\ \vec{p} = \frac{m\vec{v}}{\sqrt{1 - \frac{v^2}{c^2}}},\ \text{write}\ v^2 = \frac{p^2 c^2}{p^2 + m^2 c^2}$$

$$H = m\gamma \frac{p^2 c^2}{p^2 + m^2 c^2} + \frac{mc^2}{\gamma} \tag{10.22}$$

$$\bigg|\ \text{then,}\ \gamma = \frac{1}{\sqrt{1 - \frac{v^2}{c^2}}} = \frac{1}{\sqrt{1 - \frac{1}{c^2}\left(\frac{p^2 c^2}{p^2 + m^2 c^2}\right)}} = \frac{\sqrt{p^2 + m^2 c^2}}{mc}$$

$$H = m\frac{\sqrt{p^2 + m^2 c^2}}{mc}\frac{p^2 c^2}{p^2 + m^2 c^2} + \frac{m^2 c^3}{\sqrt{p^2 + m^2 c^2}} \tag{10.23}$$

$$H = \frac{p^2 c + m^2 c^3}{\sqrt{p^2 + m^2 c^2}} \tag{10.24}$$

$$H = c\sqrt{p^2 + m^2 c^2} \tag{10.25}$$

$$H = \sqrt{p^2 c^2 + m^2 c^4} \tag{10.26}$$

---

[3]We can ignore the first term α, as it does not contribute in the Euler-Lagrange equations.

and indeed we get the expected Hamiltonian. We are confident of the free relativistic Lagrangian $L = -mc^2\sqrt{1 - \frac{v^2}{c^2}}$ so now we shall include the interaction with an external EM field. We take inspiration from the electric potential energy $q\phi$ and the magnetic energy $\int \vec{J} \cdot \vec{A} dV = \int \rho \vec{v} \cdot \vec{A} dV$. For a particle with trajectory $\vec{r}(t)$, $\rho = q\delta^3(\vec{r} - \vec{r}(t))$, the magnetic energy is $q\vec{v} \cdot \vec{A}$. We let α_1 and α_2 be possible (dimensionless) constants, so the Lagrangian for the charged particle interacting with external EM fields should be of the form

$$L = -mc^2\sqrt{1 - \frac{v^2}{c^2}} + \alpha_1 q\phi + \alpha_2 q\vec{v} \cdot \vec{A} \tag{10.27}$$

We shall determine α_1 and α_2 by requiring the equations of motion become $\frac{d}{dt}m\vec{v}\gamma = q\vec{E} + q(\vec{v} \times \vec{B})$. So,

$$\frac{\partial L}{\partial \vec{x}} - \frac{d}{dt}\frac{\partial L}{\partial \vec{v}} = 0 \tag{10.28}$$

$$\alpha_1 q \frac{\partial \phi}{\partial \vec{x}} + \alpha_2 q \frac{\partial \vec{v} \cdot \vec{A}}{\partial \vec{x}} - \frac{d}{dt}\frac{\partial}{\partial \vec{v}}\left(-mc^2\sqrt{1 - \frac{v^2}{c^2}}\right) - \alpha_2 q \frac{d}{dt}\frac{\partial}{\partial \vec{v}}\vec{v} \cdot \vec{A} = 0 \tag{10.29}$$

recall that $\frac{\partial}{\partial \vec{v}}\left(-mc^2\sqrt{1 - \frac{v^2}{c^2}}\right) = m\vec{v}\gamma$ |

note that $\frac{\partial \phi}{\partial \vec{x}} = \vec{\nabla}\phi$ and $\frac{\partial}{\partial \vec{v}}\vec{v} \cdot \vec{A} = \vec{A}$ |

$$\alpha_1 q\vec{\nabla}\phi - \alpha_2 q\frac{d\vec{A}}{dt} - \frac{d}{dt}(m\vec{v}\gamma) + \alpha_2 q\vec{\nabla}(\vec{v} \cdot \vec{A}) = 0 \tag{10.30}$$

use $\underbrace{\underbrace{\vec{\nabla}(\vec{v} \cdot \vec{A}) = \vec{v} \times (\vec{\nabla} \times \vec{A}) + \vec{A} \times (\vec{\nabla} \times \vec{v}) + (\vec{v} \cdot \vec{\nabla})\vec{A} + (\vec{A} \cdot \vec{\nabla})\vec{v}}_{\text{vector identity}}}_{\substack{\text{2nd and 4th terms in the identity are zero} \\ \text{since } \vec{v} \text{ and } \vec{x} \text{ are treated as independent}}}$ |

write $\vec{\nabla} \times \vec{A} = \vec{B}$ and $\frac{d\vec{A}(\vec{x},t)}{dt} = \frac{\partial \vec{A}}{\partial t} + (\vec{v} \cdot \vec{\nabla})\vec{A}$ |

$$\alpha_1 q\vec{\nabla}\phi - \alpha_2 q\frac{\partial \vec{A}}{\partial t} - \alpha_2 q(\vec{v} \cdot \vec{\nabla})\vec{A} - \frac{d}{dt}(m\vec{v}\gamma) + \alpha_2 q\vec{v} \times \vec{B} + \alpha_2 q(\vec{v} \cdot \vec{\nabla})\vec{A} = 0 \tag{10.31}$$

for $\alpha_1 \vec{\nabla}\phi - \alpha_2 \frac{\partial \vec{A}}{\partial t} = \vec{E}$, we need $\alpha_1 = -1$ and $\alpha_2 = +1$ |

$$q\vec{E} - \frac{d}{dt}(m\vec{v}\gamma) + q\vec{v} \times \vec{B} = 0 \tag{10.32}$$

so, $\boxed{L = -mc^2\sqrt{1 - \frac{v^2}{c^2}} - q\phi + q\vec{v} \cdot \vec{A}} \tag{10.33}$

and to get the Hamiltonian, we first need the conjugate momentum

$$\vec{p} = \frac{\partial L}{\partial \vec{v}} = \frac{\partial}{\partial \vec{v}}\left(-mc^2\sqrt{1 - \frac{v^2}{c^2}} + q\vec{v} \cdot \vec{A}\right) = m\vec{v}\gamma + q\vec{A} \tag{10.34}$$

Then perform the Legendre transform

$$H = \vec{v} \cdot \vec{p} - L \tag{10.35}$$

$$H = \vec{v} \cdot \left(m\vec{v}\gamma + q\vec{A}\right) + mc^2\sqrt{1 - \frac{v^2}{c^2}} + q\phi - q\vec{v}\cdot\vec{A} \tag{10.36}$$

$$H = mv^2\gamma + mc^2\sqrt{1 - \frac{v^2}{c^2}} + q\phi \tag{10.37}$$

$$\mid \text{ write } m\vec{v}\gamma = \vec{p} - q\vec{A} \text{ and replace all } v^2 = \frac{(\vec{p}-q\vec{A})^2 c^2}{(\vec{p}-q\vec{A})^2 + m^2 c^2}$$

$$H = \sqrt{(\vec{p}-q\vec{A})^2 c^2 + m^2 c^4} + q\phi \tag{10.38}$$

Since we must have γL to be a Lorentz scalar, we need to check if the 2 extra terms fulfil it.

$$\gamma L = -mc^2 - q\gamma\phi + q\gamma\vec{v}\cdot\vec{A} \tag{10.39}$$

$$\gamma L = -mc^2 - q(c\gamma)\left(\frac{\phi}{c}\right) + q\gamma\vec{v}\cdot\vec{A} \tag{10.40}$$

$$\mid \text{ recall 4-velocity } u^\mu = (c\gamma, \vec{v}\gamma) \text{ and 4-potential } A^\mu = \left(\frac{\phi}{c}, \vec{A}\right)$$

$$\gamma L = -mc^2 + qu^\mu A_\mu \tag{10.41}$$

Thus, the second term is indeed a Lorentz scalar since (total) charge is Lorentz invariant.

The last thing to do in this section is to put the first term into a (manifestly) Lorentz scalar form and derive the covariant equation of motion $K^\mu = \frac{dp^\mu}{d\tau} = qF^{\mu\nu}u_\nu$. We look at only the first term (free particle term) and we also think about Hamilton's principle which is to extremise the trajectory and ask "Where is the meaning of trajectory in the action expression $S = \int (-mc^2) d\tau$?"

To answer that question, we have to go back to the invariant interval which is the trajectory (or worldline) in Minkowski space![4]

$$\text{Recall, } ds^2 = -c^2 d\tau^2 = -c^2 dt^2 + dx^2 + dy^2 + dz^2 = \eta_{\mu\nu} dx^\mu dx^\nu \tag{10.42}$$

$$ds = \sqrt{-c^2 d\tau^2} = \sqrt{-1}\, cd\tau \tag{10.43}$$

$$\text{thus, } S = -mc \int \frac{1}{\sqrt{-1}} ds \tag{10.44}$$

$$\mid \text{ but } ds = \sqrt{\eta_{\mu\nu} dx^\mu dx^\nu} = \sqrt{\eta_{\mu\nu}\frac{dx^\mu}{d\tau}\frac{dx^\nu}{d\tau}} d\tau = \sqrt{u_\mu u^\mu}\, d\tau \tag{10.45}$$

$$S = -mc \int \sqrt{\frac{u_\mu u^\mu}{-1}}\, d\tau \tag{10.46}$$

$$S = -mc \int \sqrt{-u_\mu u^\mu}\, d\tau \tag{10.47}$$

Our final invariant action is

$$\boxed{S = -mc \int \sqrt{-u_\mu u^\mu}\, d\tau + \int qu^\mu A_\mu\, d\tau = \int \tilde{L}\, d\tau} \tag{10.48}$$

---

[4] Actually we are parameterizing the worldline using the proper time as parameter. This is natural especially for particles with mass, the notion of proper time is meaningful. In more general terms, we can use some parameter λ and write $ds = \sqrt{\eta_{\mu\nu} dx^\mu dx^\nu} = \sqrt{\eta_{\mu\nu}\frac{dx^\mu}{d\lambda}\frac{dx^\nu}{d\lambda}}\, d\lambda$.

We can deduce the (covariant) Euler-Lagrange equations to be

$$\frac{\partial \tilde{L}}{\partial x_\mu} - \frac{d}{d\tau}\left(\frac{\partial \tilde{L}}{\partial\left(\frac{dx_\mu}{d\tau}\right)}\right) = 0 \qquad (10.49)$$

$$\boxed{\frac{\partial \tilde{L}}{\partial x_\mu} - \frac{d}{d\tau}\frac{\partial \tilde{L}}{\partial u_\mu} = 0} \qquad (10.50)$$

$$-\frac{d}{d\tau}\frac{\partial}{\partial u_\mu}\left(-mc\sqrt{-u_\nu u^\nu} + qu^\nu A_\nu\right) + \frac{\partial}{\partial x_\mu}(qu^\nu A_\nu) = 0 \qquad (10.51)$$

$$-\frac{d}{d\tau}\frac{mcu^\mu}{\sqrt{-u_\nu u^\nu}} - q\frac{dA^\mu}{d\tau} + qu^\nu\frac{\partial A_\nu}{\partial x_\mu} = 0 \qquad (10.52)$$

note that $u_\nu u^\nu = -c^2$ can be deduced from equation (7.123)|

$$-\frac{dp^\mu}{d\tau} - q\frac{dx^\nu}{d\tau}\frac{\partial A^\mu}{\partial x^\nu} + qu^\nu \partial^\mu A_\nu = 0 \qquad (10.53)$$

$$qu^\nu(\partial^\mu A_\nu - \partial_\nu A^\mu) = \frac{dp^\mu}{d\tau} \qquad (10.54)$$

$$qu_\nu(\partial^\mu A^\nu - \partial^\nu A^\mu) = \frac{dp^\mu}{d\tau} \qquad (10.55)$$

$$qu_\nu F^{\mu\nu} = \frac{dp^\mu}{d\tau} \qquad (10.56)$$

which gives the formula (equation (8.73)) that we had to guess earlier.

10.3 Lagrangian of the EM field

The field has infinite degrees of freedom because it has a value (or values) at every spacetime point and there are infinitely many points. Thus the Lagrangian has to be specified at every point which is uncountable. The more sensible, well-defined quantity would be the Lagrangian per unit volume or the Lagrangian density.

$$\text{Lagrangian: } L = \int \mathcal{L} d^3\vec{x} \qquad (10.57)$$

$$\text{Action: } S = \int_{t_i}^{t_f} L dt = \int_{t_i}^{t_f}\int \mathcal{L} d^3\vec{x} dt \qquad (10.58)$$

To determine if the Lagrangian density $\mathcal{L}$ is a Lorentz scalar or not, we look at how the volume element $d^3\vec{x}dt$ transforms under Lorentz transformations. To do that, we look at the Jacobian.

$$dt'dx'dy'dz' = \begin{vmatrix} \frac{\partial t'}{\partial t} & \frac{\partial t'}{\partial x} & \frac{\partial t'}{\partial y} & \frac{\partial t'}{\partial z} \\ \frac{\partial x'}{\partial t} & \frac{\partial x'}{\partial x} & \frac{\partial x'}{\partial y} & \frac{\partial x'}{\partial z} \\ \frac{\partial y'}{\partial t} & \frac{\partial y'}{\partial x} & \frac{\partial y'}{\partial y} & \frac{\partial y'}{\partial z} \\ \frac{\partial z'}{\partial t} & \frac{\partial z'}{\partial x} & \frac{\partial z'}{\partial y} & \frac{\partial z'}{\partial z} \end{vmatrix} dtdxdydz \qquad (10.59)$$

| with $t' = \gamma\left(t - \frac{u}{c^2}x\right)$, $x' = \gamma(x - ut)$, $y' = y$ and $z' = z$

$$dt'dx'dy'dz' = \begin{vmatrix} \gamma & -\gamma\frac{u}{c^2} & 0 & 0 \\ -\gamma u & \gamma & 0 & 0 \\ 0 & 0 & 1 & 0 \\ 0 & 0 & 0 & 1 \end{vmatrix} dtdxdydz \tag{10.60}$$

$$dt'dx'dy'dz' = \left(\gamma \begin{vmatrix} \gamma & 0 & 0 \\ 0 & 1 & 0 \\ 0 & 0 & 1 \end{vmatrix} + \gamma\frac{u}{c^2}\begin{vmatrix} -\gamma u & 0 & 0 \\ 0 & 1 & 0 \\ 0 & 0 & 1 \end{vmatrix}\right) dtdxdydz \tag{10.61}$$

$$dt'dx'dy'dz' = \left(\gamma^2 - \gamma^2\frac{u^2}{c^2}\right) dtdxdydz \tag{10.62}$$

$$dt'dx'dy'dz' = dtdxdydz \tag{10.63}$$

Thus, the volume element is a Lorentz scalar and so for the action to be a Lorentz scalar, the Lagrangian density $\mathcal{L}$ must be a Lorentz scalar also.

We shall "upgrade" the particle Lagrangian $L(q,\dot{q};t)$ to the field Lagrangian density $\mathcal{L}$ by the following extrapolations:

- The position variable $q(t)$ shall be replaced by the 4-potential: $q(t) \longrightarrow A_\mu(\bar{x})$. This means we assume that $A_\mu(\bar{x})$ is the fundamental variable in the Lagrangian density $\mathcal{L}$.

- The velocity variable $\dot{q}(t)$ shall be replaced by $\partial_\mu A_\nu(\bar{x})$ since now $A_\nu(\bar{x})$ depends on 4-position, so all derivatives (or "velocities") must be considered.

- So overall: $L(q(t), \dot{q}(t); t) \longrightarrow \mathcal{L}(A_\mu(\bar{x}), \partial_\mu A_\nu(\bar{x}))$

To get the field Euler-Lagrange equations of motion, we shall carry out the procedure of variational calculus,

$$\text{Field variation: } A'_\mu(\bar{x}) = A_\mu(\bar{x}) + \delta A_\mu(\bar{x}) \tag{10.64}$$

$$\text{Fixed endpoints: } \delta A_\mu(\vec{x}, t_i) = 0 = \delta A_\mu(\vec{x}, t_f) \tag{10.65}$$

$$0 = \delta S \tag{10.66}$$

$$0 = \int_{t_i}^{t_f} \int d^3\vec{x} \, dt \, \delta\mathcal{L} \tag{10.67}$$

$$0 = \int_{t_i}^{t_f} \int d^3\vec{x} \, dt \left(\frac{\partial \mathcal{L}}{\partial A_\nu}\delta A_\nu + \frac{\partial \mathcal{L}}{\partial(\partial_\mu A_\nu)}\delta(\partial_\mu A_\nu)\right) \tag{10.68}$$

| the variation and the partial derivative commutes: $\delta(\partial_\mu A_\nu) = \partial_\mu(\delta A_\nu)$

| (1 dimensional) proof: $\delta(\partial_x A_\nu) = \lim_{\epsilon \to 0} \frac{1}{\epsilon}(A'_\nu(x+\epsilon) - A'_\nu(x) - (A_\nu(x+\epsilon) - A_\nu(x)))$

$$= \lim_{\epsilon \to 0} \frac{1}{\epsilon}(A'_\nu(x+\epsilon) - A_\nu(x+\epsilon) - (A'_\nu(x) - A_\nu(x)))$$

$$= \lim_{\epsilon \to 0} \frac{1}{\epsilon}(\delta A_\nu(x+\epsilon) - \delta A_\nu(x)) = \partial_x(\delta A_\nu)$$

$$0 = \int_{t_i}^{t_f} \int d^3\vec{x} \, dt \left(\frac{\partial \mathcal{L}}{\partial A_\nu}\delta A_\nu + \frac{\partial \mathcal{L}}{\partial(\partial_\mu A_\nu)}\partial_\mu \delta A_\nu\right) \tag{10.69}$$

| integrate by parts on the second term

$$0 = \int_{t_i}^{t_f} \int d^3\vec{x} \, dt \left(\frac{\partial \mathcal{L}}{\partial A_\nu}\delta A_\nu + \partial_\mu\left(\frac{\partial \mathcal{L}}{\partial(\partial_\mu A_\nu)}\delta A_\nu\right) - \partial_\mu\left(\frac{\partial \mathcal{L}}{\partial(\partial_\mu A_\nu)}\right)\delta A_\nu\right) \tag{10.70}$$

$$0 = \int_{t_i}^{t_f} \int d^3\vec{x} \, dt \left[\frac{\partial \mathcal{L}}{\partial A_\nu} - \partial_\mu\left(\frac{\partial \mathcal{L}}{\partial(\partial_\mu A_\nu)}\right)\right]\delta A_\nu + \int d^3\vec{x} \frac{\partial \mathcal{L}}{\partial(\partial_\mu A_\nu)}\delta A_\nu \Big|_{t_i}^{t_f} \tag{10.71}$$

the last term is zero due to the fixed endpoints condition. Since δA_ν is an arbitrary variation, $\delta S = 0$ only if

$$\boxed{\frac{\partial \mathcal{L}}{\partial A_\nu} - \partial_\mu \left(\frac{\partial \mathcal{L}}{\partial (\partial_\mu A_\nu)} \right) = 0} \qquad (10.72)$$

which are the desired Euler-Lagrange equations of motion for fields. Since $\mathcal{L}$ is a Lorentz scalar, the equations of motion are Lorentz covariant equations (as they should be).

The Euler-Lagrange equations work also for fields interacting with external sources. In the case of electrodynamics with external sources, we expect to get $\partial_\nu F^{\mu\nu} = \mu_0 J^\mu$. For the free (or non-interacting) part of the Lagrangian, we only have 2 Lorentz scalars to choose from: $F^{\mu\nu} F_{\mu\nu}$ or $\widetilde{F}^{\mu\nu} F_{\mu\nu}$. We choose $F^{\mu\nu} F_{\mu\nu}$ and the reason is explained as exercise 11.1. For the interaction part, we choose the Lorentz scalar $q u_\mu A^\mu = J_\mu A^\mu$ which is the usual vector coupling term inspired from equation (10.48).

$$\mathcal{L} = \alpha_1 F^{\mu\nu} F_{\mu\nu} + \alpha_2 J_\mu A^\mu \quad \text{(where } \alpha_1 \text{ and } \alpha_2 \text{ are constants to determine)} \qquad (10.73)$$

We will now determine these constants by using the fact that the equations of motion are the Maxwell equations.

$$\frac{\partial \mathcal{L}}{\partial A_\nu} - \partial_\mu \left(\frac{\partial \mathcal{L}}{\partial (\partial_\mu A_\nu)} \right) = 0 \qquad (10.74)$$

recall that $F^{\mu\nu} = \partial^\mu A^\nu - \partial^\nu A^\mu$ |

$$\frac{\partial (\alpha_2 J^\mu A_\mu)}{\partial A_\nu} - \partial_\mu \left[\frac{\partial}{\partial (\partial_\mu A_\nu)} \alpha_1 (\partial^\sigma A^\gamma - \partial^\gamma A^\sigma)(\partial_\sigma A_\gamma - \partial_\gamma A_\sigma) \right] = 0 \qquad (10.75)$$

expand and note $\partial^\sigma A^\gamma \frac{\partial (\partial_\sigma A_\gamma)}{\partial (\partial_\mu A_\nu)} = \partial^\sigma A^\gamma \delta^\mu_\sigma \delta^\nu_\gamma = \partial^\mu A^\nu$ |

$$\alpha_2 J^\nu - \alpha_1 \partial_\mu (2\partial^\mu A^\nu - 2\partial^\nu A^\mu - 2\partial^\nu A^\mu + 2\partial^\mu A^\nu) = 0 \qquad (10.76)$$

$$\alpha_2 J^\nu - 4\alpha_1 \partial_\mu F^{\mu\nu} = 0 \qquad (10.77)$$

We want $\partial_\nu F^{\mu\nu} = \mu_0 J^\mu$ so $\alpha_1 = -\frac{1}{4\mu_0}$ and $\alpha_2 = 1$.

$$\boxed{\mathcal{L} = -\frac{1}{4\mu_0} F^{\mu\nu} F_{\mu\nu} + J_\mu A^\mu} \qquad (10.78)$$

Note a bonus identity from the above working:

$$\frac{\partial}{\partial (\partial_\mu A_\nu)} \left(-\frac{1}{4\mu_0} F^{\alpha\beta} F_{\alpha\beta} \right) = -\frac{1}{\mu_0} F^{\mu\nu} \qquad (10.79)$$

Next, we shall look at the conservation laws or Noether's theorem in field theory. As usual, we only look at 1st order (symmetry) transformations.

Coordinate transformation: $x'^\mu = x^\mu + \delta x^\mu$ $\qquad (10.80)$

Corresponding change in field: $A'_\nu(\bar{x}') = A_\nu(\bar{x}) + \delta A_\nu(\bar{x})$ $\qquad (10.81)$

Define a "field variation": $\bar{\delta} A_\nu(\bar{x}) = A'_\nu(\bar{x}) - A_\nu(\bar{x})$ $\qquad (10.82)$

which is variation of the field while keeping the coordinates fixed. This variation is useful because it commutes with the partial derivative (as we have seen earlier). The 2 variations are related by (keeping everything to 1st order),

$$\bar{\delta} A_\nu(\bar{x}) = A'_\nu(\bar{x}) - A_\nu(\bar{x}) \qquad (10.83)$$

Lagrangian Description of a Classical Relativistic $U(1)$ Gauge Theory

$$\bar{\delta}A_\nu(\bar{x}) = A'_\nu(\bar{x}) - A'_\nu(\bar{x}') + A'_\nu(\bar{x}') - A_\nu(\bar{x}) \tag{10.84}$$

$$\bar{\delta}A_\nu(\bar{x}) = -(A'_\nu(\bar{x}') - A'_\nu(\bar{x})) + \delta A_\nu(\bar{x}) \tag{10.85}$$

$$\bar{\delta}A_\nu(\bar{x}) = \delta A_\nu(\bar{x}) - (A'_\nu(\bar{x} + \delta\bar{x}) - A'_\nu(\bar{x})) \tag{10.86}$$

$$\bar{\delta}A_\nu(\bar{x}) = \delta A_\nu(\bar{x}) - \frac{\partial A'_\nu(\bar{x})}{\partial x^\mu} \delta x^\mu \tag{10.87}$$

$$\bar{\delta}A_\nu(\bar{x}) = \delta A_\nu(\bar{x}) - (\partial_\mu A'_\nu(\bar{x}))\delta x^\mu \tag{10.88}$$

| to 1st order, we can write $A'_\nu(\bar{x}) \to A_\nu(\bar{x})$

$$\bar{\delta}A_\nu(\bar{x}) = \delta A_\nu(\bar{x}) - (\partial_\mu A_\nu(\bar{x}))\delta x^\mu \tag{10.89}$$

We demand that these symmetry transformations leave the action invariant (to 1st order).

$$0 = \delta S \tag{10.90}$$

$$0 = \delta \int\int \mathcal{L} d^3\vec{x} dt \tag{10.91}$$

$$0 = \int\int (\delta\mathcal{L}) d^3\vec{x} dt + \int\int \mathcal{L}\delta(d^3\vec{x} dt) \tag{10.92}$$

| the variation of the volume element is found by calculating the Jacobian

$$d^3\vec{x}' dt' = \begin{vmatrix} \frac{\partial x'^0}{\partial x^0} & \frac{\partial x'^0}{\partial x^1} & \cdots \\ \frac{\partial x'^1}{\partial x^0} & \frac{\partial x'^1}{\partial x^1} & \cdots \\ \vdots & \vdots & \ddots \end{vmatrix} d^3\vec{x} dt = \begin{vmatrix} 1 + \frac{\partial \delta x^0}{\partial x^0} & \frac{\partial \delta x^0}{\partial x^1} & \cdots \\ \frac{\partial \delta x^1}{\partial x^0} & 1 + \frac{\partial \delta x^1}{\partial x^1} & \cdots \\ \vdots & \vdots & \ddots \end{vmatrix} d^3\vec{x} dt$$

| then $d^3\vec{x}' dt' \approx \left(1 + \frac{\partial \delta x^\mu}{\partial x^\mu}\right) d^3\vec{x} dt = (1 + \partial_\mu \delta x^\mu) d^3\vec{x} dt$ so $\delta(d^3\vec{x} dt) = (\partial_\mu \delta x^\mu) d^3\vec{x} dt$

$$0 = \int\int (\delta\mathcal{L} + \mathcal{L}(\partial_\mu \delta x^\mu)) d^3\vec{x} dt \tag{10.93}$$

| recall $\bar{\delta}A_\nu(\bar{x}) = \delta A_\nu(\bar{x}) - (\partial_\mu A_\nu(\bar{x}))\delta x^\mu$ and infer $\delta\mathcal{L} = \bar{\delta}\mathcal{L} + (\partial_\mu \mathcal{L})\delta x^\mu$

$$0 = \int\int \left(\bar{\delta}\mathcal{L} + \mathcal{L}(\partial_\mu \delta x^\mu) + (\partial_\mu \mathcal{L})\delta x^\mu\right) d^3\vec{x} dt \tag{10.94}$$

$$0 = \int\int \left(\bar{\delta}\mathcal{L} + \partial_\mu(\mathcal{L}\delta x^\mu)\right) d^3\vec{x} dt \tag{10.95}$$

| so, $\bar{\delta}\mathcal{L} = \frac{\partial \mathcal{L}}{\partial A_\mu}\bar{\delta}A_\mu + \frac{\partial \mathcal{L}}{\partial(\partial_\mu A_\nu)}\bar{\delta}(\partial_\mu A_\nu)$

| add & subtract a term, $\bar{\delta}\mathcal{L} = \frac{\partial \mathcal{L}}{\partial A_\mu}\bar{\delta}A_\mu - \partial_\mu\left(\frac{\partial \mathcal{L}}{\partial(\partial_\mu A_\nu)}\right)\bar{\delta}A_\nu$

| $\qquad\qquad\qquad\qquad + \partial_\mu\left(\frac{\partial \mathcal{L}}{\partial(\partial_\mu A_\nu)}\right)\bar{\delta}A_\nu + \frac{\partial \mathcal{L}}{\partial(\partial_\mu A_\nu)}\partial_\mu\bar{\delta}A_\nu$

| then $\bar{\delta}\mathcal{L} = \left[\frac{\partial \mathcal{L}}{\partial A_\mu} - \partial_\mu\left(\frac{\partial \mathcal{L}}{\partial(\partial_\mu A_\nu)}\right)\right]\bar{\delta}A_\nu + \partial_\mu\left(\frac{\partial \mathcal{L}}{\partial(\partial_\mu A_\nu)}\bar{\delta}A_\nu\right)$

| using Euler-Lagrange equations, the first term is zero

$$0 = \int\int \partial_\mu\left[\frac{\partial \mathcal{L}}{\partial(\partial_\mu A_\nu)}\bar{\delta}A_\nu + \mathcal{L}\delta x^\mu\right] d^3\vec{x} dt \tag{10.96}$$

| recall $\bar{\delta}A_\nu = \delta A_\nu - (\partial_\sigma A_\nu)\delta x^\sigma$

$$0 = \int\int \partial_\mu\left[\frac{\partial \mathcal{L}}{\partial(\partial_\mu A_\nu)}\delta A_\nu - \left(\frac{\partial \mathcal{L}}{\partial(\partial_\mu A_\nu)}(\partial_\sigma A_\nu) - \delta^\mu_\sigma \mathcal{L}\right)\delta x^\sigma\right] d^3\vec{x} dt \tag{10.97}$$

We are thus invited to postulate (not rigorous)[5] local conservation laws of the form $\partial_\mu J^\mu = 0$. The so-called Noether's current density is thus defined as[6]

$$J^\mu = \frac{\partial \mathcal{L}}{\partial(\partial_\mu A_\nu)} \delta A_\nu - \left(\frac{\partial \mathcal{L}}{\partial(\partial_\mu A_\nu)}(\partial_\sigma A_\nu) - \delta^\mu_\sigma \mathcal{L}\right) \delta x^\sigma \qquad (10.98)$$

Now with regards to the energy-momentum tensor we saw earlier (equation (8.107)), we shall apply Noether's theorem to the special case where the action is invariant under 4D translations,

$$x'^\mu = x^\mu + \epsilon^\mu \Longrightarrow \delta x^\mu = \epsilon^\mu \qquad (10.99)$$

and the fields do not change under the 4D translations

$$A'_\nu(\bar{x}') = A_\nu(\bar{x}) \Longrightarrow \delta A_\nu = 0 \qquad (10.100)$$

$$0 = \partial_\mu J^\mu \qquad (10.101)$$

$$0 = \partial_\mu \left[\frac{\partial \mathcal{L}}{\partial(\partial_\mu A_\nu)}(\partial_\sigma A_\nu) - \delta^\mu_\sigma \mathcal{L}\right] \epsilon^\sigma \qquad (10.102)$$

| since ϵ^σ is arbitrary,

$$0 = \partial_\mu \left[\frac{\partial \mathcal{L}}{\partial(\partial_\mu A_\nu)}(\partial_\sigma A_\nu) - \delta^\mu_\sigma \mathcal{L}\right] \qquad (10.103)$$

| raise the index σ by multiplying with $\eta^{\sigma\alpha}$

$$0 = \partial_\mu \left[\frac{\partial \mathcal{L}}{\partial(\partial_\mu A_\nu)}(\partial^\alpha A_\nu) - \eta^{\mu\alpha} \mathcal{L}\right] \qquad (10.104)$$

Thus we can define the (non-symmetric) canonical energy-momentum tensor $\Theta^{\mu\alpha}$,[7]

$$\Theta^{\mu\alpha} = \frac{\partial \mathcal{L}}{\partial(\partial_\mu A_\nu)}(\partial^\alpha A_\nu) - \eta^{\mu\alpha} \mathcal{L} \qquad (10.105)$$

Note that since we are discussing 4D translational invariance, we have to use $\mathcal{L} = \mathcal{L}_{\text{free}} = -\frac{1}{4\mu_0} F^{\mu\nu} F_{\mu\nu}$. Interaction with external sources is not translationally invariant.

The quantity that is conserved is revealed when we integrate $0 = \partial_\mu J^\mu$ over 3D space.

$$0 = \int \partial_\mu J^\mu dV \qquad (10.106)$$

$$= \int \partial_0 J^0 dV + \int \vec{\nabla} \cdot \vec{J} dV \qquad (10.107)$$

| use divergence theorem on second term

$$0 = \frac{d}{dx^0} \int J^0 dV + \oint \vec{J} \cdot d\vec{A} \qquad (10.108)$$

| assume the fields and therefore $\vec{J}$ fall off sufficiently fast

$$0 = \frac{d}{dt} \int J^0 dV \qquad (10.109)$$

---

[5] I have browsed through various classical/quantum field theory books and the authoritative [2], and I am unable to find any satisfying rigorous derivation of this.

[6] Be careful that Noether's current density and the external current density have the same notation but they may mean different quantities.

[7] Note that this energy-momentum tensor is actually not gauge invariant!

Thus $\int J^0 dV$ is the conserved quantity.[8] For the (canonical) energy-momentum tensor $\Theta^{\mu\nu}$, where $\partial_\mu \Theta^{\mu\nu} = 0$, the 4 conserved quantities are $\int \Theta^{0\nu} dV$. These 4 conserved quantities turn out to be the energy and the momentum.

$$\frac{1}{c}\int \Theta^{0\nu} dV = P^\nu = \left(\frac{E}{c}, \vec{P}\right) \tag{10.110}$$

Earlier, (equation (8.107)) we have a symmetric energy-momentum tensor $T^{\mu\nu}$ and it is actually related to $\Theta^{\mu\nu}$ by a 4-divergence.[9]

$$T^{\mu\nu} = \Theta^{\mu\nu} + \partial_\sigma \chi^{\sigma\mu\nu} \quad (\text{where we require } \chi^{\sigma\mu\nu} = -\chi^{\mu\sigma\nu}) \tag{10.111}$$

This relationship does not affect the conservation law.

$$\partial_\mu T^{\mu\nu} = \partial_\mu \Theta^{\mu\nu} + \partial_\mu \partial_\sigma \chi^{\sigma\mu\nu} \tag{10.112}$$

$\quad$ | from Noether's theorem, $\partial_\mu \Theta^{\mu\nu} = 0$

$\quad$ | split $\chi^{\sigma\mu\nu} = \frac{1}{2}(\chi^{\sigma\mu\nu} + \chi^{\sigma\mu\nu})$

$$\partial_\mu T^{\mu\nu} = \frac{1}{2}\partial_\mu \partial_\sigma \chi^{\sigma\mu\nu} + \frac{1}{2}\partial_\mu \partial_\sigma \chi^{\sigma\mu\nu} \tag{10.113}$$

$\quad$ | in second term, rename indices $\mu \to \sigma$ and $\sigma \to \mu$

$$\partial_\mu T^{\mu\nu} = \frac{1}{2}\partial_\mu \partial_\sigma \chi^{\sigma\mu\nu} + \frac{1}{2}\partial_\sigma \partial_\mu \chi^{\mu\sigma\nu} \tag{10.114}$$

$\quad$ | partial derivatives commute $\partial_\sigma \partial_\mu = \partial_\mu \partial_\sigma$

$$\partial_\mu T^{\mu\nu} = \frac{1}{2}\partial_\mu \partial_\sigma (\chi^{\sigma\mu\nu} + \chi^{\mu\sigma\nu}) \tag{10.115}$$

$\quad$ | recall the requirement $\chi^{\sigma\mu\nu} = -\chi^{\mu\sigma\nu}$

$$\partial_\mu T^{\mu\nu} = 0 \tag{10.116}$$

which is the conservation law for the non-interacting case. The conserved quantities are also not affected.

$$\int T^{0\nu} dV = \int \left(\Theta^{0\nu} + \partial_\sigma \chi^{\sigma 0\nu}\right) dV \tag{10.117}$$

$$\int T^{0\nu} dV = \int \Theta^{0\nu} dV + \int \partial_0 \chi^{00\nu} dV + \int \partial_i \chi^{i0\nu} dV \tag{10.118}$$

$\quad$ | but $\chi^{00\nu} = 0$ since $\chi^{00\nu} = -\chi^{00\nu}$

$\quad$ | use divergence theorem in the 3rd term

$\quad$ | and assume χ falls off sufficiently fast

$$\int T^{0\nu} dV = \int \Theta^{0\nu} dV \tag{10.119}$$

---

[8] A quick example is to recall the 4-current $J^\mu = (c\rho, \vec{J})$ where $J^0 = c\rho$ and so $\int J^0 dV = Q$, the total charge, is the conserved quantity.

[9] If you are interested, you can read pg 112-114 of "Electrodynamics and Classical theory of Fields and Particles" by A.O. Barut on how to get $T^{\mu\nu}$ directly from variation, rather than my method where I am "repairing $\Theta^{0\nu}$ to get $T^{\mu\nu}$".

Now we shall take the explicit (free field) expressions of $\Theta^{\mu\nu}$ and $T^{\mu\nu}$ and work out $\chi^{\sigma\mu\nu}$.

$$\Theta^{\mu\nu} = \frac{\partial \mathcal{L}}{\partial(\partial_\mu A_\sigma)}(\partial^\nu A_\sigma) - \eta^{\mu\nu}\mathcal{L} \tag{10.120}$$

| use $\mathcal{L} = \mathcal{L}_0 = -\frac{1}{4\mu_0}F_{\alpha\beta}F^{\alpha\beta}$

| and we already know from eq(10.79) $\frac{\partial \mathcal{L}_0}{\partial(\partial_\mu A_\sigma)} = -\frac{1}{\mu_0}F^{\mu\sigma}$

$$\Theta^{\mu\nu} = -\frac{1}{\mu_0}F^{\mu\sigma}(\partial^\nu A_\sigma) + \frac{1}{4\mu_0}\eta^{\mu\nu}F^{\alpha\beta}F_{\alpha\beta} \tag{10.121}$$

$$T^{\mu\nu} = -\frac{1}{\mu_0}\left[\eta^{\mu\sigma}F_{\sigma\alpha}F^{\alpha\nu} + \frac{1}{4}\eta^{\mu\nu}F^{\alpha\beta}F_{\alpha\beta}\right] \tag{10.122}$$

| replace $\frac{1}{4}\eta^{\mu\nu}F^{\alpha\beta}F_{\alpha\beta} = \Theta^{\mu\nu} + \mu_0 F^{\mu\sigma}(\partial^\nu A_\sigma)$

$$T^{\mu\nu} = -\frac{1}{\mu_0}[\eta^{\mu\sigma}F_{\sigma\alpha}F^{\alpha\nu} + F^{\mu\sigma}(\partial^\nu A_\sigma) + \mu_0\Theta^{\mu\nu}] \tag{10.123}$$

$$T^{\mu\nu} = -\frac{1}{\mu_0}[F^{\mu\alpha}F_\alpha{}^\nu + F^{\mu\alpha}(\partial^\nu A_\alpha) + \mu_0\Theta^{\mu\nu}] \tag{10.124}$$

$$T^{\mu\nu} = -\frac{1}{\mu_0}[F^{\mu\alpha}(\partial_\alpha A^\nu - \partial^\nu A_\alpha + \partial^\nu A_\alpha) + \mu_0\Theta^{\mu\nu}] \tag{10.125}$$

$$T^{\mu\nu} = -\frac{1}{\mu_0}[F^{\mu\alpha}(\partial_\alpha A^\nu) + \mu_0\Theta^{\mu\nu}] \tag{10.126}$$

| use product rule on the first term and $\partial_\alpha F^{\mu\alpha} = 0$ (for free field)

$$T^{\mu\nu} = -\frac{1}{\mu_0}[\partial_\alpha(F^{\mu\alpha}A^\nu) + \mu_0\Theta^{\mu\nu}] \tag{10.127}$$

| write $F^{\mu\alpha} = -F^{\alpha\mu}$

$$T^{\mu\nu} = -\frac{1}{\mu_0}[-\partial_\alpha(F^{\alpha\mu}A^\nu) + \mu_0\Theta^{\mu\nu}] \tag{10.128}$$

$$T^{\mu\nu} = \frac{1}{\mu_0}\partial_\alpha(F^{\alpha\mu}A^\nu) + \Theta^{\mu\nu} \tag{10.129}$$

Thus $\chi^{\alpha\mu\nu} = \frac{1}{\mu_0}F^{\alpha\mu}A^\nu$ which is indeed antisymmetric in the first 2 indices.

As a nice ending to Part III, we shall take the interacting case and work out (eq (8.108)) explicitly that $\partial_\mu T^{\mu\nu} = -F^{\nu\sigma}J_\sigma$.

$$\partial_\mu T^{\mu\nu} = -\frac{1}{\mu_0}\left[\partial_\mu(\eta^{\mu\sigma}F_{\sigma\alpha}F^{\alpha\nu}) + \frac{1}{4}\partial_\mu(\eta^{\mu\nu}F^{\alpha\beta}F_{\alpha\beta})\right] \tag{10.130}$$

$$\partial_\mu T^{\mu\nu} = -\frac{1}{\mu_0}\left[(\partial^\sigma F_{\sigma\alpha})F^{\alpha\nu} + F_{\sigma\alpha}(\partial^\sigma F^{\alpha\nu}) + \frac{1}{4}((\partial^\nu F^{\alpha\beta})F_{\alpha\beta} + F^{\alpha\beta}(\partial^\nu F_{\alpha\beta}))\right] \tag{10.131}$$

| because we can write $F^{\alpha\beta}(\partial^\nu F_{\alpha\beta}) = F_{\alpha\beta}(\partial^\nu F^{\alpha\beta})$, the last 2 terms add
| the first term $\partial^\sigma F_{\sigma\alpha} = -\mu_0 J_\alpha$ which are Maxwell equations

$$\partial_\mu T^{\mu\nu} = -\frac{1}{\mu_0}\left[-\mu_0 J_\alpha F^{\alpha\nu} + F_{\sigma\alpha}(\partial^\sigma F^{\alpha\nu}) + \frac{1}{2}F_{\alpha\beta}(\partial^\nu F^{\alpha\beta})\right] \tag{10.132}$$

| for 1st term: $J_\alpha F^{\alpha\nu} = -F^{\nu\sigma}J_\sigma$, for 2nd term: rename $\sigma \to \beta$ and split it,
| for 3rd term: rename $\alpha \leftrightarrow \beta$

Lagrangian Description of a Classical Relativistic $U(1)$ Gauge Theory

$$\partial_\mu T^{\mu\nu} = -F^{\nu\sigma}J_\sigma - \frac{1}{2\mu_0}\left[F_{\beta\alpha}(\partial^\beta F^{\alpha\nu} + \partial^\beta F^{\alpha\nu} + \partial^\nu F^{\beta\alpha})\right] \quad (10.133)$$

| use homogenous Maxwell equation $\partial^\beta F^{\alpha\nu} + \partial^\nu F^{\beta\alpha} + \partial^\alpha F^{\nu\beta} = 0$

$$\partial_\mu T^{\mu\nu} = -F^{\nu\sigma}J_\sigma - \frac{1}{2\mu_0}\left[F_{\beta\alpha}(\partial^\beta F^{\alpha\nu} - \partial^\alpha F^{\nu\beta})\right] \quad (10.134)$$

| write $F^{\nu\beta} = -F^{\beta\nu}$

$$\partial_\mu T^{\mu\nu} = -F^{\nu\sigma}J_\sigma - \frac{1}{2\mu_0}\left[F_{\beta\alpha}(\partial^\beta F^{\alpha\nu} + \partial^\alpha F^{\beta\nu})\right] \quad (10.135)$$

| terms in round brackets are symmetric in $\alpha\beta$
| but $F_{\beta\alpha}$ is antisymmetric in $\alpha\beta$ so the 2nd term vanishes

$$\partial_\mu T^{\mu\nu} = -F^{\nu\sigma}J_\sigma \quad \text{(indeed)} \quad (10.136)$$

10.3.1 Where to go after this book

Now that we have "elevated" Electrodynamics to the level of the Action Principle, let me list down some ways in which further development can take place from here.

The next logical step would be to quantise the EM field and treat it as a quantum field.

- Quantum Optics: The quantised EM field is now interacting with electronic states in atoms, molecules and solids.

- Quantum Field Theory (QFT): The quantised EM field is now interacting among other quantised fields.

  - The other quantised fields are the particle (fermionic) fields of quarks and leptons and force (bosonic) fields of the EM force, the strong nuclear force and the weak nuclear force.

  - The strong nuclear force is described by the Quantum Chromodynamics (QCD) theory.

  - Eventually we have to treat EM and weak nuclear force as a single force, the electroweak force. That is the Glashow-Weinberg-Salam (GSW) Electroweak theory.

  - The electroweak force with the strong force, i.e. the GSW theory with QCD theory is the Standard model of Particle Physics.

  - Going even further, unifying gravity with the Standard Model of Particle Physics, would require, in one instance, Supersymmetry, Supergravity and then Superstring Theory. This book is really just the very beginning of all these!

11

Exercises for Part III

11.1 Checking expressions

1. Check that $F_{\alpha\beta}F^{\alpha\beta} = -\dfrac{2\vec{E}^2}{c^2} + 2\vec{B}^2$.

2. Check that $\tilde{F}^{\mu\nu}F_{\mu\nu} = \dfrac{4}{c}\vec{E}\cdot\vec{B}$.

3. Starting with $T^{\mu\nu} = -\dfrac{1}{\mu_0}\left(\eta^{\mu\sigma}F_{\sigma\alpha}F^{\alpha\nu} + \dfrac{1}{4}\eta^{\mu\nu}F_{\alpha\beta}F^{\alpha\beta}\right)$,

   (a) Show that $T^{0i} = \dfrac{S_i}{c}$.

   (b) Show that $T^{ij} = -\epsilon_0\left(E_iE_j - \dfrac{1}{2}\delta_{ij}\vec{E}^2\right) - \dfrac{1}{\mu_0}\left(B_iB_j - \dfrac{1}{2}\delta_{ij}\vec{B}^2\right)$

4. Show that $\Theta^{\mu\nu}$ is not gauge invariant for the free EM Lagrangian density $\mathcal{L} = -\dfrac{1}{4}F^{\mu\nu}F_{\mu\nu}$. Then recall $T^{\mu\nu} = \Theta^{\mu\nu} + \partial_\sigma\chi^{\sigma\mu\nu}$ and gauge transform $\partial_\sigma\chi^{\sigma\mu\nu}$. You should realise that the term $\partial_\sigma\chi^{\sigma\mu\nu}$ restores gauge invariance to $T^{\mu\nu}$.

5. Recall Lorentz scalar formed by the field and its dual is $\tilde{F}^{\mu\nu}F_{\mu\nu} = \dfrac{4}{c}\vec{E}\cdot\vec{B}$. Now show that this Lorentz scalar is not suitable for use in a Lagrangian by showing that the action $S = \int\int \tilde{F}^{\mu\nu}F_{\mu\nu}\,d^3\vec{x}dt$ is a surface term.

Part IV

Solutions to the Problems

12

Solutions to Exercises in Part I

12.1 Solution to exercise 6.1

Recall the formula for force $\vec{F} = \oint \overleftrightarrow{T} \cdot d\vec{A} - \epsilon_0 \mu_0 \frac{d}{dt} \int \vec{S} dV$ and the second term is zero as it is a static situation here. So we only have to do the closed surface integral for the northern hemisphere. There are 2 faces in the northern hemisphere: bowl face and disk face.

Bowl Face: The electric field on the surface of the bowl is $\vec{E} = \frac{1}{4\pi\epsilon_0} \frac{Q}{R^2} \hat{r}$ and the area element is $d\vec{A} = R^2 \sin\theta d\theta d\phi \hat{r}$. We will do this in Cartesian coordinates: $\hat{r} = \sin\theta \cos\phi \hat{x} + \sin\theta \sin\phi \hat{y} + \cos\theta \hat{z}$.

$$\vec{E} = \frac{1}{4\pi\epsilon_0} \frac{Q}{R^2} (\sin\theta \cos\phi \hat{x} + \sin\theta \sin\phi \hat{y} + \cos\theta \hat{z}) \tag{12.1}$$

$$T_{ij} = \epsilon_0 \left(E_i E_j - \frac{1}{2} \delta_{ij} \vec{E}^2 \right) + \frac{1}{\mu_0} \left(B_i B_j - \frac{1}{2} \delta_{ij} \vec{B}^2 \right) \tag{12.2}$$

but as there is no magnetic field in this problem,

$$= \epsilon_0 \left(E_i E_j - \frac{1}{2} \delta_{ij} \vec{E}^2 \right) \tag{12.3}$$

As we can guess that the net force is in the z-direction, we shall only T_{xz}, T_{yz} and T_{zz}.

$$T_{xz} = \epsilon_0 (E_x E_z) = \epsilon_0 \left(\frac{Q}{4\pi\epsilon_0 R^2} \right)^2 \sin\theta \cos\phi \cos\theta \tag{12.4}$$

$$T_{yz} = \epsilon_0 (E_y E_z) = \epsilon_0 \left(\frac{Q}{4\pi\epsilon_0 R^2} \right)^2 \sin\theta \cos\theta \sin\phi \tag{12.5}$$

$$T_{zz} = \epsilon_0 \left(E_z^2 - \frac{1}{2} \vec{E}^2 \right) \tag{12.6}$$

$$= \epsilon_0 \left(E_z^2 - \frac{1}{2} E_x^2 - \frac{1}{2} E_y^2 - \frac{1}{2} E_z^2 \right) \tag{12.7}$$

$$= \frac{\epsilon_0}{2} \left(E_z^2 - E_x^2 - E_y^2 \right) \tag{12.8}$$

$$= \frac{\epsilon_0}{2} \left(\frac{Q}{4\pi\epsilon_0 R^2} \right)^2 (\cos^2\theta - \sin^2\theta \cos^2\phi - \sin^2\theta \sin^2\phi) \tag{12.9}$$

$$= \frac{\epsilon_0}{2} \left(\frac{Q}{4\pi\epsilon_0 R^2} \right)^2 (\cos^2\theta - \sin^2\theta) \tag{12.10}$$

DOI: 10.1201/9781003515210-12

$$(\overleftrightarrow{T} \cdot d\vec{A})_z = T_{zx}dA_x + T_{zy}dA_y + T_{zz}dA_z \tag{12.11}$$

$$= \epsilon_0 \left(\frac{Q}{4\pi\epsilon_0 R^2}\right)^2 \Big[\sin\theta\cos\theta\cos\phi R^2 \sin^2\theta\cos\phi$$

$$+ \sin\theta\cos\theta\sin\phi R^2 \sin^2\theta\sin\phi + \frac{1}{2}(\cos^2\theta - \sin^2\theta)R^2 \sin\theta\cos\theta\Big] d\theta d\phi$$

| using trigo identity $\sin^2 + \cos^2 = 1$

$$= \frac{\epsilon_0}{2}\left(\frac{Q}{4\pi\epsilon_0 R}\right)^2 \sin\theta\cos\theta \, d\theta d\phi \tag{12.12}$$

In z-direction:
$$F_{\text{bowl}} = \oint (\overleftrightarrow{T} \cdot d\vec{A})_z \tag{12.13}$$

$$= \frac{\epsilon_0}{2}\left(\frac{Q}{4\pi\epsilon_0 R}\right)^2 \int_0^{2\pi} d\phi \int_0^{\pi/2} \sin\theta\cos\theta d\theta \tag{12.14}$$

$$= \frac{1}{4\pi\epsilon_0}\frac{Q^2}{8R^2} \tag{12.15}$$

Disk Face: Inside the sphere at the disk, the electric field is $\vec{E} = \frac{1}{4\pi\epsilon_0}\frac{Qr}{R^3}\hat{r} = \frac{1}{4\pi\epsilon_0}\frac{Qr}{R^3}(\cos\phi\hat{x} + \sin\phi\hat{y})$ in polar coordinates. The area vector is $d\vec{A} = r dr d\phi(-\hat{z})$.

$$(\overleftrightarrow{T} \cdot d\vec{A})_z = T_{zz}dA_z \tag{12.16}$$

$$= \frac{\epsilon_0}{2}\left(E_z^2 - E_x^2 - E_y^2\right)(-r dr d\phi) \tag{12.17}$$

$$= \frac{\epsilon_0}{2}\left(\frac{Q}{4\pi\epsilon_0 R^3}\right)^2 r^3(\cos^2\phi + \sin^2\phi) dr d\phi \tag{12.18}$$

$$F_{\text{disk}} = \frac{\epsilon_0}{2}\left(\frac{Q}{4\pi\epsilon_0 R^3}\right)^2 \int_0^R dr r^3 \int_0^{2\pi} d\phi \tag{12.19}$$

$$= \frac{1}{4\pi\epsilon_0}\frac{Q^2}{16R^2} \tag{12.20}$$

Finally, the net force:

$$\vec{F}_{\text{net}} = \left(\frac{1}{4\pi\epsilon_0}\frac{Q^2}{8R^2} + \frac{1}{4\pi\epsilon_0}\frac{Q^2}{16R^2}\right)\hat{z} \tag{12.21}$$

$$= \frac{1}{4\pi\epsilon_0}\frac{3Q^2}{16R^2}\hat{z} \tag{12.22}$$

12.2 Solution to exercise 6.2

1. Coulomb gauge is $\vec{\nabla} \cdot \vec{A}' = 0$.

$$\vec{\nabla} \cdot \vec{A}' = 0 \tag{12.23}$$

insert $\vec{A}' = \vec{A} + \vec{\nabla}\lambda$ |

$$\vec{\nabla} \cdot \vec{A} + \vec{\nabla}^2 \lambda = 0 \tag{12.24}$$

$$\vec{\nabla}^2 \lambda = -\vec{\nabla} \cdot \vec{A} \tag{12.25}$$

and this is like a "Poisson Equation" and so we shall borrow the solution of Poisson Equation. If we assume the boundary condition $\lim_{r \to \infty} \lambda(\vec{r}, t) = 0$ then the

Solutions to Exercises in Part I

solution of the gauge function is the familiar

$$\lambda(\vec{r},t) = \frac{1}{4\pi} \int \frac{(\vec{\nabla}\cdot\vec{A})(\vec{r}',t)}{|\vec{r}-\vec{r}'|} d^3\vec{r}' \qquad (12.26)$$

2. Lorenz gauge is $\vec{\nabla}\cdot\vec{A}' + \epsilon_0\mu_0 \frac{\partial \phi'}{\partial t} = 0$.

$$\vec{\nabla}\cdot\vec{A}' + \epsilon_0\mu_0 \frac{\partial \phi'}{\partial t} = 0 \qquad (12.27)$$

$$\Big| \text{ insert } \vec{A}' = \vec{A} + \vec{\nabla}\lambda \text{ and } \phi' = \phi - \frac{\partial \lambda}{\partial t}$$

$$0 = \vec{\nabla}\cdot(\vec{A} + \vec{\nabla}\lambda) + \epsilon_0\mu_0\left(\frac{\partial \phi}{\partial t} - \frac{\partial^2}{\partial t^2}\right)\lambda \qquad (12.28)$$

$$\left(\vec{\nabla}^2 - \epsilon_0\mu_0 \frac{\partial^2}{\partial t^2}\right)\lambda = -\left(\vec{\nabla}\cdot\vec{A} + \epsilon_0\mu_0 \frac{\partial \phi}{\partial t}\right) \qquad (12.29)$$

$$\Big| \text{ compare with equation: } \left(\vec{\nabla}^2 - \epsilon_0\mu_0 \frac{\partial^2}{\partial t^2}\right)\phi = -\frac{\rho}{\epsilon_0}$$

$$\Big| \text{ whose solution is: } \phi(\vec{r},t) = \frac{1}{4\pi\epsilon_0}\int\int \frac{\rho(\vec{r}',t')\delta\left(t-t'-\frac{|\vec{r}-\vec{r}'|}{c}\right)}{|\vec{r}-\vec{r}'|} d^3\vec{r}' dt'$$

$$\Big| \text{ so immediately the solution for } \lambda(\vec{r},t) \text{ is}$$

$$\lambda(\vec{r},t) = \frac{1}{4\pi}\int\int \frac{(\vec{\nabla}\cdot\vec{A} + \epsilon_0\mu_0 \frac{\partial\phi}{\partial t})(\vec{r}',t')}{|\vec{r}-\vec{r}'|} \delta\left(t-t'-\frac{|\vec{r}-\vec{r}'|}{c}\right) d^3\vec{r}' dt' \qquad (12.30)$$

So for $\bar{\lambda} = \lambda + f$ to satisfy the same equation as λ means,

$$\left(\vec{\nabla}^2 - \epsilon_0\mu_0 \frac{\partial^2}{\partial t^2}\right)(\bar{\lambda} - f) = -\left(\vec{\nabla}\cdot\vec{A} + \epsilon_0\mu_0 \frac{\partial \phi}{\partial t}\right) \qquad (12.31)$$

which means

$$\left(\vec{\nabla}^2 - \epsilon_0\mu_0 \frac{\partial^2}{\partial t^2}\right)f = 0 \qquad (12.32)$$

which is the homogeneous wave equation and so f is any general plane wave solution which is a linear superposition of plane waves.

12.3 Solution to exercise 6.3

We start by calculating towards the expression $\vec{\nabla}^2\phi(\vec{r},t)$

$$\vec{\nabla}\phi(\vec{r},t) = \frac{1}{4\pi\epsilon_0}\int\left[\frac{1}{|\vec{r}-\vec{r}'|}(\vec{\nabla}\rho) + \rho\left(\vec{\nabla}\frac{1}{|\vec{r}-\vec{r}'|}\right)\right]d^3\vec{r}' \qquad (12.33)$$

$$\Big| \text{ use chain rule } \vec{\nabla}\rho = (\vec{\nabla}t_r)\frac{\partial \rho}{\partial t_r} = \left(-\frac{1}{c}\vec{\nabla}R\right)\frac{\partial \rho}{\partial t} = -\frac{1}{c}\hat{R}\dot{\rho}$$

$$\Big| \text{ then } \vec{\nabla}\frac{1}{|\vec{r}-\vec{r}'|} = \vec{\nabla}\frac{1}{R} = -\frac{\hat{R}}{R^2}$$

$$= \frac{1}{4\pi\epsilon_0}\int\left[-\frac{\dot{\rho}\hat{R}}{cR} - \rho\frac{\hat{R}}{R^2}\right]d^3\vec{r}' \qquad (12.34)$$

then,

$$\vec{\nabla}^2\phi = \vec{\nabla}\cdot(\vec{\nabla}\phi) \qquad (12.35)$$

| use identity: $\vec{\nabla}\cdot(f\vec{A}) = f(\vec{\nabla}\cdot\vec{A}) + \vec{A}\cdot(\vec{\nabla}f)$

$$= \frac{1}{4\pi\epsilon_0}\int\left\{-\frac{1}{c}\left[\frac{\hat{R}}{R}\cdot(\vec{\nabla}\dot{\rho}) + \dot{\rho}\left(\vec{\nabla}\cdot\frac{\hat{R}}{R}\right)\right] - \left[\frac{\hat{R}}{R^2}\cdot(\vec{\nabla}\rho) + \rho\left(\vec{\nabla}\cdot\frac{\hat{R}}{R^2}\right)\right]\right\}d^3\vec{r}'$$

| use chain rule to get $\vec{\nabla}\dot{\rho} = -\frac{1}{c}\hat{R}\ddot{\rho}$

| use the identities: $\vec{\nabla}\cdot\frac{\hat{R}}{R} = \frac{1}{R^2}$ and $\vec{\nabla}\cdot\frac{\hat{R}}{R^2} = 4\pi\delta^3(\vec{R})$

$$= \frac{1}{4\pi\epsilon_0}\int\left[\frac{1}{c^2}\frac{\ddot{\rho}}{R} - \frac{1}{c}\frac{\dot{\rho}}{R^2} + \frac{1}{c}\frac{\dot{\rho}}{R^2} - \rho 4\pi\delta^3(\vec{r}-\vec{r}')\right]d^3\vec{r}' \qquad (12.36)$$

| recognise that the first term is 2-time derivatives of ϕ

$$= \frac{1}{c^2}\frac{\partial^2\phi}{\partial t^2} - \frac{1}{\epsilon_0}\int\rho(\vec{r}',t_r)\delta^3(\vec{r}-\vec{r}')d^3\vec{r}' \qquad (12.37)$$

| since $t_r|_{\vec{r}=\vec{r}'} = \left(t - \frac{|\vec{r}-\vec{r}'|}{c}\right)_{\vec{r}=\vec{r}'} = t$

$$= \frac{1}{c^2}\frac{\partial^2\phi}{\partial t^2} - \frac{\rho(\vec{r},t)}{\epsilon_0} \qquad (12.38)$$

which indeed satisfies the retarded potential equation.

12.4 Solution to exercise 6.4

We start with $\vec{\nabla}\cdot\vec{A}(\vec{r},t)$,

$$\vec{\nabla}\cdot\vec{A}(\vec{r},t) = \frac{\mu_0}{4\pi}\int\left(\vec{\nabla}\cdot\frac{\vec{J}}{R}\right)d^3\vec{r}' \qquad (12.39)$$

| so use $\vec{\nabla}\cdot\frac{\vec{J}}{R} = \frac{1}{R}(\vec{\nabla}\cdot\vec{J}) + \vec{J}\cdot\left(\vec{\nabla}\frac{1}{R}\right)$

| then use $\vec{\nabla}\frac{1}{R} = -\vec{\nabla}'\frac{1}{R}$

$$= \frac{\mu_0}{4\pi}\int\left[\frac{1}{R}(\vec{\nabla}\cdot\vec{J}) - \vec{J}\cdot\left(\vec{\nabla}'\frac{1}{R}\right)\right] \qquad (12.40)$$

| then reverse the product rule for the second term

$$= \frac{\mu_0}{4\pi}\int\left[\frac{1}{R}(\vec{\nabla}\cdot\vec{J}) - \vec{\nabla}'\cdot\left(\frac{\vec{J}}{R}\right) + \frac{1}{R}(\vec{\nabla}'\cdot\vec{J})\right]d^3\vec{r}' \qquad (12.41)$$

| then $\vec{\nabla}\cdot\vec{J} = \frac{\partial J_x}{\partial x} + \cdots = \frac{\partial J_x}{\partial t_r}\frac{\partial t_r}{\partial x} + \cdots = \frac{\partial J_x}{\partial t_r}\frac{\partial\left(t-\frac{R}{c}\right)}{\partial x} + \cdots$

$$= -\frac{1}{c}\frac{\partial J_x}{\partial t_r}\frac{\partial R}{\partial x} + \cdots = -\frac{1}{c}\frac{\partial\vec{J}}{\partial t_r}\cdot(\vec{\nabla}R)$$

| then $\vec{\nabla}'\cdot\vec{J}(\vec{r}',t_r) = \underbrace{\vec{\nabla}'\cdot\vec{J}}_{\text{explicit }\vec{r}'} - \underbrace{\frac{1}{c}\frac{\partial\vec{J}}{\partial t_r}\cdot(\vec{\nabla}'R)}_{\text{implicit }\vec{r}'}$

| continuity equation is used on explicit $\vec{\nabla}' \cdot \vec{J}$
| to give $\vec{\nabla}' \cdot \vec{J}(\vec{r}', t_r) = -\dfrac{\partial \rho}{\partial t} - \dfrac{1}{c}\dfrac{\partial \vec{J}}{\partial t_r} \cdot (\vec{\nabla}' R)$

$$= \frac{\mu_0}{4\pi} \int \left[-\frac{1}{Rc}\frac{\partial \vec{J}}{\partial t_r} \cdot (\vec{\nabla} R) - \left(\vec{\nabla}' \cdot \frac{\vec{J}}{R}\right) - \frac{1}{R}\frac{\partial \rho}{\partial t} - \frac{1}{Rc}\frac{\partial \vec{J}}{\partial t_r} \cdot (\vec{\nabla}' R) \right] d^3 \vec{r}'$$

| 1st and 4th term cancels as $\vec{\nabla} R = -\vec{\nabla}' R$
| 2nd term becomes surface term using divergence theorem
| then assume it falls off to zero so net flux is zero

$$= -\frac{\mu_0}{4\pi} \int \frac{1}{R}\frac{\partial \rho}{\partial t} d^3 \vec{r}' \tag{12.42}$$

$$= -\frac{\mu_0}{4\pi} \frac{4\pi\epsilon_0}{4\pi\epsilon_0} \frac{\partial}{\partial t} \int \frac{\rho}{R} d^3 r' \tag{12.43}$$

$$= -\epsilon_0 \mu_0 \frac{\partial \phi}{\partial t} \tag{12.44}$$

12.5 Solution to exercise 6.5

Start[1] with the Jefimenko equation for $\vec{B}$:

$$\vec{B}(\vec{r}, t) = \frac{\mu_0}{4\pi} \int \left[\frac{\vec{J}(\vec{r}', t_r)}{R^2} + \frac{\dot{\vec{J}}(\vec{r}', t_r)}{cR} \right] \times \hat{R}\ d^3 \vec{r}' \tag{12.49}$$

| for 1st term, $\vec{J}(\vec{r}', t_r) \approx \vec{J}(\vec{r}', t) + (t_r - t)\dot{\vec{J}}(\vec{r}', t)$
| for 2nd term, $\dot{\vec{J}}(\vec{r}', t_r) \approx \dot{\vec{J}}(\vec{r}', t)$ to the same order

$$\approx \frac{\mu_0}{4\pi} \int \left[\frac{\vec{J}(\vec{r}', t)}{R^2} + \frac{(t_r - t)}{R^2}\dot{\vec{J}}(\vec{r}', t) + \frac{\dot{\vec{J}}(\vec{r}', t)}{cR} \right] \times \hat{R}\ d^3 \vec{r}' \tag{12.50}$$

| recall that $t_r = t - \dfrac{R}{c}$
| then 2nd and 3rd terms cancel

$$= \frac{\mu_0}{4\pi} \int \frac{\vec{J}(\vec{r}', t) \times \hat{R}}{R^2} d^3 \vec{r}' \tag{12.51}$$

---

[1] We drop the argument $\vec{r}'$ for better clarity in the derivation.

$$\vec{J}(t_r) = \vec{J}(t) + (t_r - t)\dot{\vec{J}}(t) + \cdots \tag{12.45}$$

| more explicitly,

$$= \vec{J}(t_r)\big|_{t_r = t} + \frac{\partial \vec{J}(t_r)}{\partial t_r}\bigg|_{t_r = t} (t_r - t) + \cdots \tag{12.46}$$

| take $\dfrac{\partial}{\partial t_r}$ on both sides

$$\frac{\partial \vec{J}(t_r)}{\partial t_r} = \frac{\partial \vec{J}(t_r)}{\partial t_r}\bigg|_{t_r = t} \tag{12.47}$$

$$\Rightarrow \dot{\vec{J}}(t_r) = \dot{\vec{J}}(t) \tag{12.48}$$

which is Biot-Savart's law. The 2nd term is the (first order) effect of retarded time and the 3rd term is the (first order) effect of the time-dependent current density. They cooperate and cancel at first order. This makes Biot-Savart's law "good" for zeroth and first order.

12.6 Solution to exercise 6.6

We start with the expression,

$$\vec{E} = \frac{q}{4\pi\epsilon_0} \frac{1 - \frac{v^2}{c^2}}{\left(1 - \frac{v^2}{c^2}\sin^2\alpha\right)^{3/2}} \frac{\hat{r}_p}{r_p^2} = \frac{q}{4\pi\epsilon_0} \frac{1 - \frac{v^2}{c^2}}{\left(1 - \frac{v^2}{c^2}\sin^2\alpha\right)^{3/2}} \frac{\vec{r}_p}{r_p^3} \quad (12.52)$$

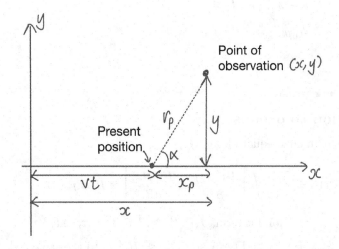

1. Finding the component parallel to the velocity. Using $\vec{r}_p = x_p\hat{x} + y\hat{y}$

$$E^\parallel = E_x = \frac{q}{4\pi\epsilon_0} \frac{1 - \frac{v^2}{c^2}}{\left(1 - \frac{v^2}{c^2}\sin^2\alpha\right)^{3/2}} \frac{x_p}{r_p^3} \quad (12.53)$$

$$\Big| \text{ let } \gamma = \frac{1}{\sqrt{1 - \frac{v^2}{c^2}}}$$

$$= \frac{q}{4\pi\epsilon_0} \frac{x_p}{\gamma^2 \left(r_p^2 - \frac{v^2}{c^2} r_p^2 \sin^2\alpha\right)^{3/2}} \quad (12.54)$$

$$\Big| \text{ then use } r_p^2 = x_p^2 + y^2 \text{ and } r_p^2 \sin^2\alpha = y^2$$

$$= \frac{q}{4\pi\epsilon_0} \frac{x_p}{\gamma^2 \left(x_p^2 + y^2 - \frac{v^2}{c^2} y^2\right)^{3/2}} \quad (12.55)$$

$$= \frac{q}{4\pi\epsilon_0} \frac{x_p}{\gamma^2 \left(x_p^2 + y^2 \frac{1}{\gamma^2}\right)^{3/2}} \quad (12.56)$$

$$= \frac{q}{4\pi\epsilon_0} \frac{\gamma x_p}{\left(x_p^2 \gamma^2 + y^2\right)^{3/2}} \quad (12.57)$$

$$= \frac{q}{4\pi\epsilon_0} \frac{\gamma(x - vt)}{\left(\gamma^2(x - vt)^2 + y^2\right)^{3/2}} \quad (12.58)$$

Solutions to Exercises in Part I

For turning point,

$$0 = \frac{dE^{\|}}{dvt} = \frac{q\gamma}{4\pi\epsilon_0} \frac{2\gamma^2(x-vt)^2 - y^2}{(\gamma^2(x-vt)^2 + y^2)^{5/2}} \quad (12.59)$$

$$\Longrightarrow 2\gamma^2(x-vt)^2 - y^2 = 0 \quad (12.60)$$

$$(x-vt)^2 = \frac{y^2}{2\gamma^2} \quad (12.61)$$

$$vt = x \mp \frac{y}{\gamma\sqrt{2}} \quad (12.62)$$

For extreme $E^{\|}$ values,

$$E^{\|}\bigg|_{(x-vt)^2 = \frac{y^2}{2\gamma^2}} = \frac{q}{4\pi\epsilon_0} \frac{\gamma\left(\pm\frac{y}{\gamma\sqrt{2}}\right)}{\left(\gamma^2 \frac{y^2}{2\gamma^2} + y^2\right)^{3/2}} \quad (12.63)$$

$$= \frac{q}{4\pi\epsilon_0} \frac{\pm\frac{y}{\sqrt{2}}}{\left(\frac{3}{2}\right)^{3/2} y^3} \quad (12.64)$$

$$= \pm\frac{q}{4\pi\epsilon_0} \sqrt{\frac{4}{27}} \frac{1}{y^2} \quad (12.65)$$

which does not depend on the speed of the particle!

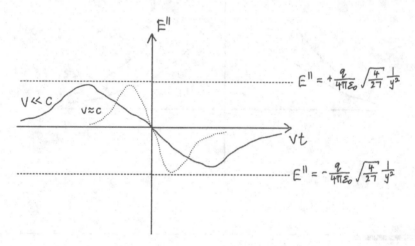

2. Finding the component perpendicular to the velocity.

$$E^{\perp} = E_y = \frac{q}{4\pi\epsilon_0} \frac{1 - \frac{v^2}{c^2}}{\left(1 - \frac{v^2}{c^2}\sin^2\alpha\right)^{3/2}} \frac{y}{r_p^3} \quad (12.66)$$

$$\bigg| \text{ similarly let } \gamma = \frac{1}{\sqrt{1 - \frac{v^2}{c^2}}}$$

$$= \frac{q}{4\pi\epsilon_0} \frac{y}{\gamma^2 \left(r_p^2 - \frac{v^2}{c^2} r_p^2 \sin^2\alpha\right)^{3/2}} \quad (12.67)$$

$$\bigg| \text{ we skip similar steps}$$

$$= \frac{q}{4\pi\epsilon_0} \frac{\gamma y}{(\gamma^2(x-vt)^2 + y^2)^{3/2}} \quad (12.68)$$

For turning point,

$$0 = \frac{dE^\perp}{dvt} = \frac{q\gamma}{4\pi\epsilon_0} \frac{3\gamma^3(x-vt)^2}{(\gamma^2(x-vt)^2+y^2)^{5/2}} \quad (12.69)$$

$$\implies vt = x \quad (12.70)$$

thus there is one peak. Now for the extreme $E^\perp$ value.

$$E^\perp\big|_{vt=x} = \frac{q}{4\pi\epsilon_0} \frac{\gamma y}{(y^2)^{3/2}} = \frac{q}{4\pi\epsilon_0} \frac{\gamma}{y^2} \quad (12.71)$$

The ratio is,

$$\frac{E^\parallel}{E^\perp} = \frac{x-vt}{y} \quad (12.72)$$

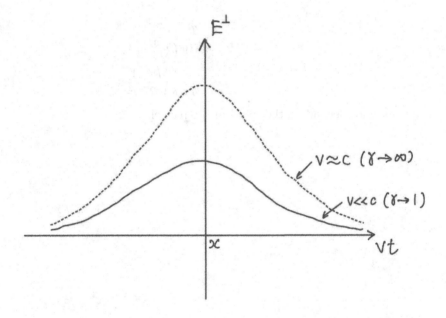

12.7 Solution to exercise 6.7

Recall Larmor's formula for a (slow moving) charged particle: $P = \frac{\mu_0 q^2 a^2}{6\pi c}$. First we need to justify that the tangential velocity is indeed small enough for Larmor's formula to be applicable.

$$\frac{mv_{\tan}^2}{r} = \frac{1}{4\pi\epsilon_0}\frac{e^2}{r^2} \implies v_{\tan} = \sqrt{\frac{e^2}{m4\pi\epsilon_0 r}} = 0.0075\,c \quad (12.73)$$

so indeed $v_{\tan} \ll c$ and we can use Larmor's formula for this. Substituting the centripetal acceleration $a = \frac{v_{\tan}^2}{r}$ into Larmor's formula,

$$P = \frac{\mu_0 e^2}{6\pi c}\left(\frac{v_{\tan}^2}{r}\right)^2 \quad (12.74)$$

Solutions to Exercises in Part I

$$\bigg| \quad \text{substituting } v_{\tan} = \sqrt{\frac{e^2}{m 4\pi\epsilon_0 r}}$$

$$= \frac{\mu_0 e^2}{6\pi c} \left(\frac{e^2}{m 4\pi\epsilon_0 r^2} \right)^2 \tag{12.75}$$

The total initial energy is $U = \frac{1}{2} m v_{\tan}^2 - \frac{e^2}{4\pi\epsilon_0 r} = \frac{1}{2} m \frac{e^2}{m 4\pi\epsilon_0 r} - \frac{e^2}{4\pi\epsilon_0 r} = -\frac{1}{2} \frac{1}{4\pi\epsilon_0} \frac{e^2}{r}$. This energy is removed at a rate of Larmor's power.

$$\frac{dU}{dt} = -P \tag{12.76}$$

$$\frac{1}{2} \frac{1}{4\pi\epsilon_0} \frac{e^2}{r^2} \frac{dr}{dt} = -\frac{\mu_0 e^2}{6\pi c} \left(\frac{e^2}{m 4\pi\epsilon_0 r^2} \right)^2 \tag{12.77}$$

$$\frac{dr}{dt} = -\frac{e^4}{m^2 12\pi^2 c^3 \epsilon_0^2} \frac{1}{r^2} \tag{12.78}$$

$$t = -\frac{12\pi^2 c^3 \epsilon_0^2 m^2}{e^4} \int_{5 \times 10^{-11}}^{0} r^2 \, dr \tag{12.79}$$

$$= \frac{4\pi^2 c^3 \epsilon_0^2 m^2}{e^4} (5 \times 10^{-11})^3 \tag{12.80}$$

$$= 1.314 \times 10^{-11} \text{ seconds} \tag{12.81}$$

12.8 Solution to exercise 6.8

Recall the Lorenz gauge is $\vec{\nabla} \cdot \vec{A} = -\epsilon_0 \mu_0 \frac{\partial \phi}{\partial t}$.

$$\text{RHS} = -\epsilon_0 \mu_0 \frac{\partial \phi}{\partial t} \tag{12.82}$$

$$= -\frac{\mu_0 p_0 \cos\theta}{4\pi r} \left[-\frac{\omega^2}{c} \cos\left(\omega\left(t - \frac{r}{c}\right)\right) - \frac{\omega}{r} \sin\left(\omega\left(t - \frac{r}{c}\right)\right) \right] \tag{12.83}$$

$$= \frac{\mu_0 p_0 \cos\theta \, \omega}{4\pi r} \left[\frac{\omega}{c} \cos\left(\omega\left(t - \frac{r}{c}\right)\right) + \frac{1}{r} \sin\left(\omega\left(t - \frac{r}{c}\right)\right) \right] \tag{12.84}$$

$$\text{LHS} = \vec{\nabla} \cdot \vec{A} \tag{12.85}$$

$$= \frac{\partial A_z}{\partial z} \tag{12.86}$$

$$= -\frac{\mu_0 p_0 \omega}{4\pi} \frac{\partial}{\partial z} \left(\frac{\sin\left(\omega\left(t - \frac{r}{c}\right)\right)}{r} \right) \tag{12.87}$$

$$= -\frac{\mu_0 p_0 \omega}{4\pi} \left[-\frac{\omega \cos\left(\omega\left(t - \frac{r}{c}\right)\right)}{c} \frac{\partial r}{\partial z} - \frac{\sin\left(\omega\left(t - \frac{r}{c}\right)\right)}{r^2} \frac{\partial r}{\partial z} \right] \tag{12.88}$$

$$\bigg| \quad \text{use } \frac{\partial r}{\partial z} = \frac{\partial}{\partial z} \sqrt{x^2 + y^2 + z^2} = \frac{z}{r} = \cos\theta$$

$$= \frac{\mu_0 p_0 \omega \cos\theta}{4\pi r} \left[\frac{\omega}{c} \cos\left(\omega\left(t - \frac{r}{c}\right)\right) + \frac{1}{r} \sin\left(\omega\left(t - \frac{r}{c}\right)\right) \right] \tag{12.89}$$

Hence RHS = LHS and Lorenz gauge is satisfied despite the approximations. This means the approximations do not spoil the consistency of the theory.

12.9 Solution to exercise 6.9

1. We start with the definition of the dipole moment from Part I, the section on "Arbitrary Distribution".

$$\vec{p} = \int \vec{r}' \rho(\vec{r}', \tilde{t}) \, d^3\vec{r}' \qquad (12.90)$$

| recall the dipole charge density is

$$q_0 \cos(\omega t_r^+) \delta^3 \left(\vec{r}' - \frac{d}{2}\hat{z} \right) + (-q_0) \cos(\omega t_r^-) \delta^3 \left(\vec{r}' + \frac{d}{2}\hat{z} \right)$$

| we need to approximate the charge density to perfect dipole as well
| so the approximation $d \ll r$ amounts to $t_r^+, t_r^- \approx t - \frac{r}{c} = \tilde{t}$
| we cannot approximate the delta function
| since d and $|\vec{r}'|$ are equally small
| so, $\rho(\vec{r}', t) \approx q_0 \cos(\omega \tilde{t}) \delta^3 \left(\vec{r}' - \frac{d}{2}\hat{z} \right) + (-q_0) \cos(\omega \tilde{t}) \delta^3 \left(\vec{r}' + \frac{d}{2}\hat{z} \right)$
| carry out the integrals using the delta functions

$$\approx \frac{d}{2}\hat{z} q_0 \cos\left(\omega\left(t - \frac{r}{c}\right)\right) - \frac{d}{2}\hat{z}(-q_0) \cos\left(\omega\left(t - \frac{r}{c}\right)\right) \qquad (12.91)$$

| let $p_0 = q_0 d$

$$\vec{p} = p_0 \cos\left(\omega\left(t - \frac{r}{c}\right)\right) \hat{z} \qquad (12.92)$$

$$\implies \ddot{\vec{p}} = -p_0 \omega^2 \cos\left(\omega\left(t - \frac{r}{c}\right)\right) \hat{z} \qquad (12.93)$$

For total power, we then have

$$P = \frac{\mu_0}{6\pi c} \ddot{p}^2 \qquad (12.94)$$

$$= \frac{\mu_0 p_0^2 \omega^4}{6\pi c} \cos^2\left(\omega\left(t - \frac{r}{c}\right)\right) \qquad (12.95)$$

The time-averaged total power is

$$\langle P \rangle_{\text{time}} = \frac{1}{T} \int_0^T P \, dt \qquad (12.96)$$

| where $T = \frac{2\pi}{\omega}$

$$= \frac{\mu_0 p_0^2 \omega^4}{6\pi c} \underbrace{\frac{1}{T} \int_0^T \cos^2\left(\omega\left(t - \frac{r}{c}\right)\right) dt}_{=\frac{1}{2}} \qquad (12.97)$$

$$= \frac{\mu_0 p_0^2 \omega^4}{12\pi c} \qquad (12.98)$$

which is the same as in the main text.

2. For point charge, the dipole moment is $\vec{p} = q\vec{r}$ where $\vec{r}$ is the position vector of the point charge. We align the (instantaneous) position vector $\vec{r}$ along the z-axis and we obtain $\ddot{\vec{p}} = qa\hat{z}$. The total power is thus $P = \frac{\mu_0}{6\pi c}\ddot{p} = \frac{\mu_0 q^2 a^2}{6\pi c}$ which is the same as in the main text.

12.10 Solution to exercise 6.10

This is simply the working in page 451 of [4].

12.11 Solution to exercise 6.11

The quantum version of Thomson scattering is Compton scattering. The differential cross section of Compton scattering is the Klein-Nishina formula.

$$\frac{d\sigma}{d\Omega} = \alpha^2 \frac{\hbar^2}{4m^2c^2} \frac{\omega'^2}{\omega^2}\left[\frac{\omega'}{\omega} + \frac{\omega}{\omega'} + 4\sin^2\theta - 2\right] \quad (12.99)$$

simply have $\frac{\omega'}{\omega} \to 1$ and $\frac{\omega}{\omega'} \to 1$

$$\approx \frac{\mu_0^2 c^2 q^4}{16\pi^2 \hbar^2} \frac{\hbar^2}{4m^2 c^2}\left[1 + 1 + 4\sin^2\theta - 2\right] \quad (12.100)$$

$$= \frac{\mu_0^2 q^4}{16\pi^2 m^2}\sin^2\theta \quad (12.101)$$

which is indeed the non-relativistic Thomson scattering cross section. The quantum part (see where the $\hbar$ is) is really in the relation between ω and ω', i.e. the scattering cross section is dependent on frequency.

12.12 Solution to exercise 6.12

1. Recall that the incident plane wave is $\vec{E}_i = E_0 e^{ikx} e^{-i\omega t}\hat{z}$ and the (far field) scattered wave is $\vec{E}_s = E_0 \frac{e^{ikr}}{r} \vec{F}(\vartheta) e^{-i\omega t}$.

   The point is that we will equate $\vec{E}_s$ with $\vec{E}(r,\theta,t) = \frac{\mu_0 \ddot{p}(t)}{4\pi}\frac{\sin\theta}{r}\hat{\theta}$ but there are 2 things to note: (i) $\vec{E}_s$ is the complex representation but $\vec{E}(r,\theta,t)$ is actually real; (ii) the θ in $\vec{E}(r,\theta,t)$ is measured from z-axis but ϑ in $\vec{F}(\vartheta)$ is measured from x-axis.

We need $\ddot{\vec{p}}(\tilde{t})$ for this system.

$$\ddot{\vec{p}}(\tilde{t}) = \left.\frac{d^2}{dt^2}(q|\vec{r}|)\right|_{t=\tilde{t}} \tag{12.102}$$

| recall the displacement is $\vec{r} = \frac{q}{m}\Re\left(\frac{\vec{E}_i}{\omega_0^2 - \omega^2 - i\gamma\omega}\right)$

$$= \frac{q^2}{m}\Re\left(\frac{-\omega^2 E_0 e^{ikx} e^{-i\omega\tilde{t}}}{\omega_0^2 - \omega^2 - i\gamma\omega}\right) \tag{12.103}$$

Now we shall equate the fields but really we can only say that the real parts are equal.

$$\Re(\vec{E}_s) = \frac{\mu_0 \ddot{\vec{p}}(\tilde{t})}{4\pi} \frac{\sin\theta}{r}\hat{\theta} \tag{12.104}$$

$$\Re\left(E_0 \frac{e^{ikr}}{r}\vec{F}(\vartheta)e^{-i\omega t}\right) = \frac{\mu_0}{4\pi}\frac{q^2}{m}\Re\left(\frac{-\omega^2 E_0 e^{ikx} e^{-i\omega\tilde{t}}}{\omega_0^2 - \omega^2 - i\gamma\omega}\right)\frac{\sin\theta}{r}\hat{\theta} \tag{12.105}$$

We require the complex expression of $\vec{F}(\vartheta)$ for the optical theorem calculation. We shall assume that we can equate the complex numbers instead of just the real part.[2]

$$E_0 \frac{e^{ikr}}{r}\vec{F}(\vartheta)e^{-i\omega t} = -\frac{\mu_0}{4\pi}\frac{q^2}{m}\frac{\omega^2 E_0 e^{ikx} e^{-i\omega\tilde{t}}}{\omega_0^2 - \omega^2 - i\gamma\omega}\frac{\sin\theta}{r}\hat{\theta} \tag{12.106}$$

| now recall $\tilde{t} = t - \frac{r}{c}$, so $e^{-i\omega\tilde{t}} = \underbrace{e^{-i\omega\left(t-\frac{r}{c}\right)} = e^{-i\omega t}e^{ikr}}_{\text{since } \omega = kc}$

$$\vec{F}(\vartheta) = -\frac{\mu_0 q^2 \omega^2}{4\pi m}\frac{e^{ikx}\sin\theta}{\omega_0^2 - \omega^2 - i\gamma\omega}\hat{\theta} \tag{12.107}$$

| in the forward direction, $\vartheta = 0$ or $\theta = \pi/2$ and $\hat{\theta} = -\hat{z}$
| scatterer is at $x = 0$

$$\vec{F}(\text{forward}) = \vec{F}(\theta = \frac{\pi}{2}) = \frac{\mu_0 q^2 \omega^2}{4\pi m}\frac{1}{\omega_0^2 - \omega^2 - i\gamma\omega}\hat{z} \tag{12.108}$$

Now[3] applying optical theorem,

$$\sigma_t = \frac{4\pi}{k}\Im\left[\hat{z}\cdot\vec{F}(\text{forward})\right] \tag{12.109}$$

$$= \frac{4\pi}{k}\frac{\mu_0 q^2 \omega^2}{4\pi m}\Im\left[\frac{1}{\omega_0^2 - \omega^2 - i\gamma\omega}\right] \tag{12.110}$$

| use $\omega = kc$

$$= \frac{\mu_0 q^2 \omega c}{m}\frac{\gamma\omega}{(\omega_0^2 - \omega^2)^2 + (\gamma\omega)^2} \tag{12.111}$$

---

[2] The proper way is to have the dipole radiation worked out in the complex representation then we equate the complex expressions. So we are quite lucky here that the assumption turns out to be correct.

[3] Actually in the main text, we can also set $x = 0$ but it is not needed as quantities in the lecture notes are squared quantities.

Solutions to Exercises in Part I

2. The dissipative force is the "damping force". This gives the total power removed from the incident wave and hence it is related to the total cross section.

$$\langle P_{\text{total}} \rangle_{\text{time}} = \left\langle \left| \vec{F}_{\text{diss}} \cdot \dot{\vec{r}} \right| \right\rangle_{\text{time}} \qquad (12.112)$$

$$\text{where } \vec{F}_{\text{diss}} = -m\gamma \frac{d\vec{r}}{dt} = -m\gamma \dot{\vec{r}}$$

$$= m\gamma \left\langle \dot{\vec{r}} \cdot \dot{\vec{r}} \right\rangle_{\text{time}} \qquad (12.113)$$

$$= m\gamma \left(\frac{q}{m}\right)^2 \left\langle \Re\left(\frac{\dot{\vec{E}}_i}{\omega_0^2 - \omega^2 - i\gamma\omega}\right) \cdot \Re\left(\frac{\dot{\vec{E}}_i}{\omega_0^2 - \omega^2 - i\gamma\omega}\right) \right\rangle_{\text{time}}$$

use complex identity $\langle \Re(\vec{A}) \cdot \Re(\vec{B}) \rangle_{\text{time}} = \frac{1}{2}\vec{A} \cdot \vec{B}^*$

$$= m\gamma \left(\frac{q}{m}\right)^2 \frac{1}{2} \frac{\dot{\vec{E}}_i \cdot \dot{\vec{E}}_i^*}{(\omega_0^2 - \omega^2 - i\gamma\omega)(\omega_0^2 - \omega^2 + i\gamma\omega)} \qquad (12.114)$$

then $\dot{\vec{E}}_i = -i\omega E_0 e^{ikx} e^{-i\omega t} \hat{z} = -i\omega \vec{E}_i$ so $\dot{\vec{E}}_i \cdot \dot{\vec{E}}_i^* = \omega^2 |E_0|^2$

$$= \frac{q^2 \gamma}{2m} \frac{\omega^2 |E_0|^2}{(\omega_0^2 - \omega^2)^2 + (\gamma\omega)^2} \qquad (12.115)$$

recall the equation (5.38): $\langle P \rangle_{\text{time}} = \frac{|E_0|^2}{2\mu_0 c} \sigma$

$$\Rightarrow \sigma_t = \frac{\mu_0 q^2 c}{m} \frac{\gamma \omega^2}{(\omega_0^2 - \omega^2)^2 + (\gamma\omega)^2} \qquad (12.116)$$

Thus, from $\sigma_t = \sigma_a + \sigma_s$, we can conclude that the power removed from incident wave is either absorbed or scattered.

12.13   Solution to exercise 6.13

It is nice that the induced dipole moment is parallel to the applied field, also, $\lambda \gg a$, means we can borrow the results of dipole radiation from Part I.

$$\frac{d\sigma}{d\Omega} = r^2 \frac{\frac{1}{r^2} \frac{dP}{d\Omega}}{\frac{|E_0|^2}{2\mu_0 c}} \qquad (12.117)$$

$$= \frac{2\mu_0 c}{|E_0|^2} \frac{dP}{d\Omega} \qquad (12.118)$$

recall the arbitrary distribution dipole radiation: $\frac{dP}{d\Omega} = \frac{\mu_0 \ddot{p}(\tilde{t})^2}{16\pi^2 c} \sin^2 \theta$

we shall also take time average

$$= \frac{2\mu_0 c}{|E_0|^2} \frac{\mu_0}{16\pi^2 c} \langle \ddot{p}(\tilde{t})^2 \rangle_{\text{time}} \sin^2 \theta \qquad (12.119)$$

so $\vec{p}(\tilde{t}) = 4\pi\epsilon_0 \frac{\epsilon_2 - \epsilon_1}{\epsilon_2 + 2\epsilon_1} a^3 \vec{E}_i(\tilde{t})$ with $\vec{E}_i = E_0 e^{ikx} e^{-i\omega \tilde{t}} \hat{z}$

then $\ddot{\vec{p}}(\tilde{t}) = -\omega^2 \vec{p}(\tilde{t})$

then $\langle \ddot{\vec{p}}(\tilde{t})^2 \rangle_{\text{time}} = \langle \Re(\ddot{\vec{p}}(\tilde{t})) \cdot \Re(\ddot{\vec{p}}(\tilde{t})) \rangle_{\text{time}} = \frac{1}{2} \ddot{\vec{p}} \cdot \ddot{\vec{p}}^* = \frac{\omega^4}{2} \left(4\pi\epsilon_0 \frac{\epsilon_2 - \epsilon_1}{\epsilon_2 + 2\epsilon_1}\right)^2 a^6 |E_0|^2$

$$\begin{aligned}
& \quad \Big| \quad \text{then use } \omega = kc \\
&= k^4 a^6 \left(\frac{\epsilon_2 - \epsilon_1}{\epsilon_2 + 2\epsilon_1}\right)^2 \sin^2\theta & (12.120) \\
\sigma &= \int_0^{2\pi} d\phi \int_0^{\pi} d\theta\, k^4 a^6 \left(\frac{\epsilon_2 - \epsilon_1}{\epsilon_2 + 2\epsilon_1}\right)^2 \sin^3\theta & (12.121) \\
&= \frac{8\pi}{3} k^4 a^6 \left(\frac{\epsilon_2 - \epsilon_1}{\epsilon_2 + 2\epsilon_1}\right)^2 & (12.122)
\end{aligned}$$

13

Solutions to Exercises for Part II

13.1 Solution to exercise 9.1

The 2 other continuous transformations are 3D rotations (3 parameters) and 4D translations (4 parameters).

The 2 discrete transformations are time reversal denoted by:

$$\text{Time reversal} = \begin{pmatrix} -1 & & & \\ & 1 & & \\ & & 1 & \\ & & & 1 \end{pmatrix} \tag{13.1}$$

and

$$\text{Parity} = \begin{pmatrix} 1 & & & \\ & -1 & & \\ & & -1 & \\ & & & -1 \end{pmatrix} \tag{13.2}$$

The Poincare group is thus made up 10 parameters: 3 boosts, 3 rotations and 4 translations.

13.2 Solution to exercise 9.2

1. (a) We set $\Delta s^2 = -1$ and the locus is a hyperbola as can be seen from:

$$-1 = -c^2 \Delta t^2 + \Delta x^2 \tag{13.3}$$

let the event 1 be at origin $(ct_1, x_1) = (0, 0)$

$$-1 = -c^2 t_2^2 + x_2^2 \tag{13.4}$$

$$\frac{(ct_2)^2}{1^2} - \frac{x_2^2}{1^2} = 1 \tag{13.5}$$

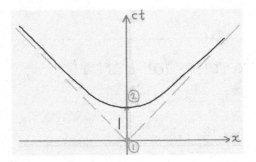

FIGURE 13.1
The locus is a hyperbola. Also called the invariant hyperbola.

(b) See figure below.

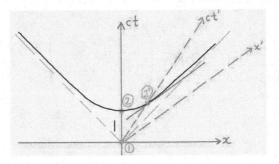

FIGURE 13.2
For question 2bii.

- Event 1 and event 2' occur at the same spatial point in S'-frame.
- The tangent at event 2' is parallel to the x-axis. In other words, points on the tangent line are simultaneous with event 2'. The gradient of the tangent line is also the velocity of that inertial frame.

2. (a) See figure 13.3 on the following pages.

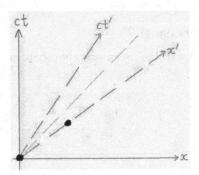

FIGURE 13.3
Frame where 2 events are simultaneous in t'.

The 2 points are spacelike separated and the dotted line S' frame is where the 2 events are simultaneous (in t'). However, for all frames, the ct' axis can't be less than 45° so we can't put both events on the ct' axis means there is no frame where the 2 events are at the same space point.

(b) See figure 13.4 on the following pages.

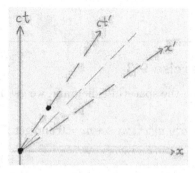

FIGURE 13.4
Frame where 2 events are at the same x'.

The 2 points are timelike separated and the dotted line S' frame is where the 2 events are at the space point (in x'). However, for all frames, the x' axis can't be more than 45° so we can't put both events on the x' axis means there is no frame where the 2 events are simultaneous (in t').

3. See Figure 13.5 on the following pages.

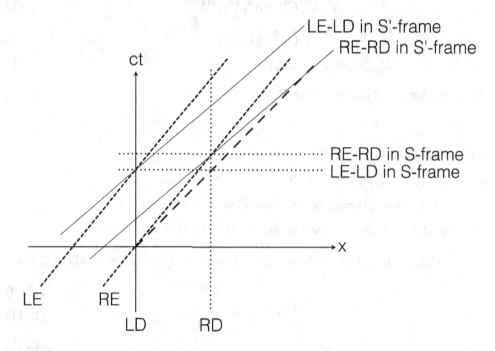

FIGURE 13.5
This is the spacetime diagram for the pole and barn paradox.

The sequence of events seen by different frames are different. In S-frame, event LE-LD occurs then RE-RD occurs. In S'-frame, RE-RD occurs then LE-LD occurs. So it is true that the different observers see different sequences of events

13.3 Solution to exercise 9.3

1. Following a 45° line on the spacetime diagram, we get $x^\mu = (ct, \vec{x}) = (9 \times 10^8, 3 \times 10^8)$.
2. Starting with $E^2 = \vec{p}^2 c^2 + m^2 c^4$, we set $m = 0$ and get $E = |\vec{p}|c$, thus $p^\mu = \left(\frac{E}{c}, \frac{E}{c}\right)$.

13.4 Solution to exercise 9.4

Since E_z is the $\mu = 0$, $\nu = 3$ element, we set $\mu = 0$ and $\nu = 3$.

$$F'^{03} = \Lambda^0{}_\alpha \Lambda^3{}_\beta F^{\alpha\beta} \tag{13.6}$$

$$\quad |\text{ expand the double sum with non-zero terms}$$

$$\frac{E'_z}{c} = \Lambda^0{}_3 \Lambda^3{}_3 F^{03} + \Lambda^0{}_1 \Lambda^3{}_3 F^{13} \tag{13.7}$$

$$\frac{E'_z}{c} = \gamma \frac{E_z}{c} + \left(-\frac{u}{c}\gamma\right)(-B_y) \tag{13.8}$$

$$E'_z = \gamma(E_z + u B_y) \tag{13.9}$$

Similarly, set $\mu = 3$ and $\nu = 1$ for B_y.

13.5 Solution to exercise 9.5

1. Since Ampere–Maxwell law involves $\vec{J}$, we need to set $\mu = 1, 2, 3$.
2. The other 2 Maxwell equations are $\vec{\nabla} \cdot \vec{B} = 0$ and $\vec{\nabla} \times \vec{E} = -\frac{\partial \vec{B}}{\partial t}$.

    - For $\vec{\nabla} \cdot \vec{B} = 0$, since there is no time in it, we shall only use indices 1, 2, 3.

    $$\partial^1 F^{23} + \partial^2 F^{31} + \partial^3 F^{12} = 0 \tag{13.10}$$

    $$\frac{\partial}{\partial x} B_x + \frac{\partial}{\partial y} B_y + \frac{\partial}{\partial z} B_z = 0 \tag{13.11}$$

    $$\vec{\nabla} \cdot \vec{B} = 0 \tag{13.12}$$

    - For $\vec{\nabla} \times \vec{E} = -\frac{\partial \vec{B}}{\partial t}$, we need to 1 time index and 2 space indices, so we have 3 types of terms which are the 3 components of the equation.

    $$\partial^0 F^{12} + \partial^1 F^{20} + \partial^2 F^{01} = 0 \tag{13.13}$$

$$\frac{1}{c}\frac{\partial}{\partial t}B_z + \frac{\partial}{\partial x}\left(-\frac{E_y}{c}\right) + \frac{\partial}{\partial y}\left(\frac{E_x}{c}\right) = 0 \qquad (13.14)$$

$$\left(\vec{\nabla}\times\vec{E}\right)_z = -\frac{\partial B_z}{\partial t} \qquad (13.15)$$

Working out the other 2 terms will give the other 2 components.

13.6 Solution to exercise 9.6

We simply work it out by setting $\mu = 0$:

$$K^0 = qF^{0\nu}u_\nu \qquad (13.16)$$

$$\frac{dp^0}{d\tau} = q\left(F^{01}u_1 + F^{02}u_2 + F^{03}u_3\right) \qquad (13.17)$$

$$\frac{1}{c}\frac{dE}{d\tau} = q\left(\frac{\vec{E}}{c}\cdot\gamma\vec{u}\right) \qquad (13.18)$$

$$|\quad \text{write } dt = \gamma d\tau$$

$$\frac{dE}{dt} = q\vec{E}\cdot\vec{u} = \vec{F}\cdot\vec{u} \qquad (13.19)$$

Thus it is power of the charged particle.

13.7 Solution to exercise 9.7

Simply work it out:

$$F'^{\mu\nu} = \partial^\mu A'^\nu - \partial^\nu A'^\mu \qquad (13.20)$$

$$|\quad \text{substitute } A'^\mu = A^\mu + \partial^\mu\lambda$$

$$= \partial^\mu A^\nu - \partial^\nu A^\mu + \partial^\mu\partial^\nu\lambda - \partial^\nu\partial^\mu\lambda \qquad (13.21)$$

$$|\quad \text{since partial derivatives can be swapped, the last 2 terms cancel}$$

$$= F^{\mu\nu} \qquad (13.22)$$

13.8 Solution to exercise 9.8

1. Substitute in the Lorentz transformation of the fields and work through the algebra:

$$\vec{B}'^2 - \frac{\vec{E}'^2}{c^2} = B'^2_x + B'^2_y + B'^2_z - \frac{E'^2_x + E'^2_y + E'^2_z}{c^2} \qquad (13.23)$$

$$|\quad \text{substitute the expressions for the Lorentz transformation}$$
$$\text{of the fields}$$

$$= B_x^2 + \gamma^2 \left(B_y + \frac{u}{c^2}E_z\right)^2 + \gamma^2 \left(B_z - \frac{u}{c^2}E_y\right)^2$$
$$- \frac{1}{c^2}E_x^2 - \frac{\gamma^2}{c^2}(E_y - uB_z)^2 - \frac{\gamma^2}{c^2}(E_z + uB_y)^2 \qquad (13.24)$$

| work out the algebra and many terms cancel

$$= B_x^2 + B_y^2 + B_z^2 - \frac{E_x^2 + E_y^2 + E_z^2}{c^2} \qquad (13.25)$$

$$= \vec{B}^2 - \frac{\vec{E}^2}{c^2} \qquad (13.26)$$

$$\vec{E}' \cdot \vec{B}' = E_x' B_x' + E_y' B_y' + E_z' B_z' \qquad (13.27)$$

| substitute the Lorentz transformations of the fields

$$= E_x B_x + \gamma(E_y - uB_z)\gamma\left(B_y + \frac{u}{c^2}E_z\right) + \gamma(E_z + uB_y)\gamma\left(B_z - \frac{u}{c^2}E_y\right)$$

| expand the terms and recall $\gamma = \dfrac{1}{\sqrt{1 - \frac{u^2}{c^2}}}$

$$= E_x B_x + E_y B_y + E_z B_z \qquad (13.28)$$
$$= \vec{E} \cdot \vec{B} \qquad (13.29)$$

2. - From $\vec{B} \cdot \vec{E} = 0$, we have $\vec{B} \perp \vec{E}$. We shall choose S'-frame to move in the x-direction (as usual). Then in the S-frame, we choose $\vec{E} = (0, E_y, 0)$ and $\vec{B} = (0, 0, B_z)$ such that $B_z^2 - \frac{E_y^2}{c^2} > 0$ (to fulfill $\vec{B}^2 - \frac{\vec{E}^2}{c^2} > 0$). Then recall the Lorentz transformed E_y: $E_y' = \gamma(E_y - uB_z)$, we get $E_y' = 0$ when $u = \frac{E_y}{B_z}$. Now $B_z^2 - \frac{E_y^2}{c^2} > 0 \implies c^2 > \left(\frac{E_y}{B_z}\right)^2 \implies c^2 > u^2$ so this S'-frame that makes $E_y' = 0$ exists.

    - We shall choose S'-frame to move in the x-direction (as usual). Then in the S-frame, we choose $\vec{E} = (0, E_y, 0)$ and $\vec{B} = (0, 0, B_z)$ such that $B_z^2 - \frac{E_y^2}{c^2} < 0$ (to fulfill $\vec{B}^2 - \frac{\vec{E}^2}{c^2} < 0$). Then recall the Lorentz transformed B_z: $B_z' = \gamma\left(B_z - \frac{u}{c^2}E_y\right)$, we get $B_z' = 0$ when $u = \frac{B_z c^2}{E_y}$. Now $B_z^2 - \frac{E_y^2}{c^2} < 0 \implies \frac{B_z^2}{E_y^2} < \frac{1}{c^2} \implies \frac{u^2}{c^4} < \frac{1}{c^2} \implies u^2 < c^2$ so this S'-frame that makes $B_z' = 0$ exists.

    - For both $B_z^2 - \frac{E_y^2}{c^2} = 0$ and $\vec{E} \cdot \vec{B} = 0$, we thus have $u = c$ and since no inertial frame can travel at the speed of light, there is no inertial frame where both E and B are zero.

3. The characteristic equation is

$$\det|F^{\mu\nu} - \lambda \eta^{\mu\nu}| = 0 \qquad (13.30)$$

| after tedious algebra

$$\lambda^4 + \lambda^2\left(\vec{B}^2 - \frac{\vec{E}^2}{c^2}\right) - \frac{4}{c}(\vec{E} \cdot \vec{B}) = 0 \qquad (13.31)$$

Eigenvalues λ represent invariant quantities of the "operator" (or matrix) $F^{\mu\nu}$. These eigenvalues are solved by a polynomial of order 4, but there are only 2 eigenvalues since only even powers appear in this equation. The coefficients are proportional to the 2 invariants so the eigenvalues are made up of these 2 invariants only.

13.9 Solution to exercise 9.9

Recall that $t_r = t - \frac{R}{c}$ or $R = c(t - t_r)$. Then the denominator,

$$Rc - \vec{R} \cdot \vec{v} = c^2(t - t_r) - \vec{R} \cdot \vec{v} \tag{13.32}$$

$$= \frac{1}{\gamma}[\gamma c(ct - ct_r) - (\vec{r} - \vec{r}_0) \cdot \gamma \vec{v}] \tag{13.33}$$

| note that the 4-velocity is $u^\nu = (\gamma c, \gamma \vec{v})$
| and note that the 4-position is $x^\mu - x_0^\mu = (ct - ct_r, \vec{r} - \vec{r}_0)$

$$= -\frac{1}{\gamma}(x - x_0(\tau))_\nu u^\nu \bigg|_{\tau - \tau_r} \tag{13.34}$$

so,
$$A^0 = \frac{\phi}{c} = -\frac{q}{4\pi\epsilon_0 c} \frac{\gamma c}{(x - x_0(\tau))_\nu u^\nu}\bigg|_{\tau-\tau_r} \tag{13.35}$$

$$= -\frac{q}{4\pi\epsilon_0} \frac{u_0}{(x - x_0(\tau))_\nu u^\nu}\bigg|_{\tau-\tau_r} \tag{13.36}$$

$$\vec{A} = -\frac{q}{4\pi\epsilon_0 c^2} \frac{c\vec{v}(t')\gamma}{(x - x_0(\tau))_\nu u^\nu}\bigg|_{\tau-\tau_r} \tag{13.37}$$

$$= -\frac{q}{4\pi\epsilon_0 c} \frac{\vec{u}}{(x - x_0(\tau))_\nu u^\nu}\bigg|_{\tau-\tau_r} \tag{13.38}$$

and the 4-potential is $A^\mu = (A^0, \vec{A})$.

13.10 Solution to exercise 9.10

The condition for maximum is (let $\beta = \frac{v}{c}$)

$$0 = \frac{d}{d\theta}\left[\frac{\sin^2\theta}{(1 - \beta\cos\theta)^5}\right] \tag{13.39}$$

$$0 = \frac{2\sin\theta\cos\theta(1 - \beta\cos\theta) - 5\sin^2\theta(\beta\sin\theta)}{(1 - \beta\cos\theta)^6} \tag{13.40}$$

$$0 = 2\sin\theta\cos\theta(1 - \beta\cos\theta) - 5\beta\sin^3\theta \tag{13.41}$$

$$0 = 3\beta\cos^2\theta + 2\cos\theta - 5\beta \tag{13.42}$$

| solving the quadratic equation, we get

$$\cos\theta = \frac{-2 \pm \sqrt{2^2 - 4(3\beta)(-5\beta)}}{6\beta} \tag{13.43}$$

| when $v \to 0$, we want a symmetrical donut, so $\theta \to 90°$
| thus $\cos\theta \to 0$ means we choose the positive sign

$$\theta_{\max} = \cos^{-1}\left(\frac{\sqrt{1+15\beta^2}-1}{3\beta}\right) \tag{13.44}$$

Now for $v \approx c$ or $\beta \approx 1$, we write $\beta = 1 - \epsilon$ and perform binomial expansion to first order only.

$$\frac{\sqrt{1+15\beta^2}-1}{3\beta} = \frac{\sqrt{1+15(1-\epsilon)^2}-1}{3(1-\epsilon)} \tag{13.45}$$

$$\approx \frac{1}{3}(1+\epsilon)\left(\sqrt{1+15(1-2\epsilon)}-1\right) \tag{13.46}$$

$$= \frac{1}{3}(1+\epsilon)\left(4\sqrt{1-\frac{30\epsilon}{16}}-1\right) \tag{13.47}$$

$$\approx \frac{1}{3}(1+\epsilon)\left(4\left(1-\frac{15\epsilon}{16}\right)-1\right) \tag{13.48}$$

$$\approx \frac{1}{3}(1+\epsilon)\left(3-\frac{15\epsilon}{4}\right) \tag{13.49}$$

$$\approx \frac{1}{3}\left(3-\frac{15\epsilon}{4}+3\epsilon\right) \tag{13.50}$$

$$= 1 - \frac{\epsilon}{4} \tag{13.51}$$

Then $\cos\theta_{\max} \approx 1 - \frac{1}{2}\theta_{\max}^2$ means $1 - \frac{1}{2}\theta_{\max}^2 = 1 - \frac{\epsilon}{4}$ and so $\theta_{\max} = \sqrt{\frac{\epsilon}{2}} = \sqrt{\frac{1}{2}(1-\beta)}$.

14

Solutions to Exercises for Part III

14.1 Solution to exercise 11.1

1. The matrix $F_{\mu\nu}$ is derived from $F_{\mu\nu} = \eta_{\mu\alpha}\eta_{\nu\beta}F^{\alpha\beta}$, so

$$F_{\mu\nu} = \begin{pmatrix} 0 & -E_x/c & -E_y/c & -E_z/c \\ E_x/c & 0 & B_z & -B_y \\ E_y/c & -B_z & 0 & B_x \\ E_z/c & B_y & -B_x & 0 \end{pmatrix} \quad (14.1)$$

then, $\quad F_{\mu\nu}F^{\mu\nu} = -\dfrac{2\vec{E}^2}{c^2} + 2\vec{B}^2 \quad (14.2)$

2. The matrix $\widetilde{F}_{\mu\nu}$ is derived from $\widetilde{F}_{\mu\nu} = \tfrac{1}{2}\epsilon_{\mu\nu\alpha\beta}F^{\alpha\beta}$, so

$$\widetilde{F}_{\mu\nu} = \begin{pmatrix} 0 & B_x & B_y & B_z \\ -B_x & 0 & E_z/c & -E_y/c \\ -B_y & -E_z/c & 0 & E_x/c \\ -B_z & E_y/c & -E_x/c & 0 \end{pmatrix} \quad (14.3)$$

then, $\quad \widetilde{F}_{\mu\nu}F^{\mu\nu} = \dfrac{4}{c}\vec{E}\cdot\vec{B} \quad (14.4)$

3. (a) Let $\mu = 0$ and $\nu = i$, so

$$T^{0i} = -\dfrac{1}{\mu_0}\left(\eta^{0\alpha}F_{\alpha\beta}F^{\beta i} + \dfrac{1}{4}\eta^{0i}F_{\alpha\beta}F^{\alpha\beta}\right) \quad (14.5)$$

| the second term is zero since $\eta^{0i} = 0$
| for first term, $\alpha = 0$ and $\eta^{00} = -1$

$$= \dfrac{1}{\mu_0}F_{0\beta}F^{\beta i} \quad (14.6)$$

$$= \dfrac{1}{\mu_0}\left(\dfrac{\vec{E}}{c}\times\vec{B}\right)_i \quad (14.7)$$

$$= \dfrac{S_i}{c} \quad (14.8)$$

(b) First we set $\mu = i$, $\nu = j$ and substitute $F_{\alpha\beta}F^{\alpha\beta} = -\frac{2\vec{E}^2}{c^2} + 2\vec{B}^2$, so

$$T^{ij} = -\frac{1}{\mu_0}\left(\eta^{i\alpha}F_{\alpha\beta}F^{\beta j} + \frac{1}{4}\eta^{ij}\left(-\frac{2\vec{E}^2}{c^2} + 2\vec{B}^2\right)\right) \quad (14.9)$$

| note that $\eta^{ij} = \delta^{ij}$ and δ^{ij} and δ_{ij} are the same
| note that we can write $\eta^{i\alpha} = \eta^{ik} = \delta^{ik}$

$$= -\frac{1}{\mu_0}\left(\eta^{ik}F_{k\beta}F^{\beta j} + \frac{1}{4}\delta_{ij}\left(-\frac{2\vec{E}^2}{c^2} + 2\vec{B}^2\right)\right) \quad (14.10)$$

$$= -\frac{1}{\mu_0}\left(F_{i\beta}F^{\beta j} - \frac{1}{2}\delta_{ij}\frac{\vec{E}^2}{c^2} + \frac{1}{2}\delta_{ij}\vec{B}^2\right) \quad (14.11)$$

$$= -\frac{1}{\mu_0}\left(F_{i0}F^{0j} + F_{i1}F^{1j} + F_{i2}F^{2j} + F_{i3}F^{3j} - \frac{1}{2}\delta_{ij}\frac{\vec{E}^2}{c^2} + \frac{1}{2}\delta_{ij}\vec{B}^2\right)$$

$$= -\frac{1}{\mu_0}\left(\frac{E_i}{c}\frac{E_j}{c} + F_{i1}F^{1j} + F_{i2}F^{2j} + F_{i3}F^{3j} - \frac{1}{2}\delta_{ij}\frac{\vec{E}^2}{c^2} + \frac{1}{2}\delta_{ij}\vec{B}^2\right)$$

| recall $\frac{1}{\mu_0 c^2} = \epsilon_0$

$$= -\epsilon_0\left(E_i E_j - \frac{1}{2}\delta_{ij}\vec{E}^2\right) - \frac{1}{\mu_0}\left(F_{i1}F^{1j} + F_{i2}F^{2j} + F_{i3}F^{3j} + \frac{1}{2}\delta_{ij}\vec{B}^2\right)$$

| the 3 terms, $F_{i1}F^{1j} + F_{i2}F^{2j} + F_{i3}F^{3j}$ shall be done case-by-case
| if $i = 1$, $j = 1$, $F_{i1}F^{1j} = 0$
| and $F_{12}F^{2j} + F_{13}F^{3j} = (B_z)(B_z) + (-B_y)(B_y) = -B_z^2 - B_y^2$
| so for $i, j = 1$, $F_{i1}F^{1j} + F_{i2}F^{2j} + F_{i3}F^{3j} + \frac{1}{2}\delta_{ij}\vec{B}^2$

$$= \frac{1}{2}B_x^2 - \frac{1}{2}B_y^2 - \frac{1}{2}B_z^2 = B_x^2 - \frac{1}{2}\vec{B}^2$$

| if $i = 1$, $j = 2$, $F_{i1}F^{1j} + F_{i2}F^{2j} + F_{i3}F^{3j} = F_{13}F^{32}$

$$= (-B_y)(-B_x) = B_x B_y$$

| all cases are consistent with $F_{i1}F^{1j} + F_{i2}F^{2j} + F_{i3}F^{3j} + \frac{1}{2}\delta_{ij}\vec{B}^2$

$$= B_i B_j - \frac{1}{2}\delta_{ij}\vec{B}^2$$

$$= -\epsilon_0\left(E_i E_j - \frac{1}{2}\delta_{ij}\vec{E}^2\right) - \frac{1}{\mu_0}\left(B_i B_j - \frac{1}{2}\delta_{ij}\vec{B}^2\right) \quad (14.12)$$

4. We recall the canonical energy-momentum tensor,

$$\Theta^{\mu\nu} = \frac{\partial \mathcal{L}}{\partial(\partial_\mu A_\sigma)}(\partial^\nu A_\sigma) - \eta^{\mu\nu}\mathcal{L} \quad (14.13)$$

| use the free field Lagrangian $\mathcal{L} = -\frac{1}{4\mu_0}F^{\mu\nu}F_{\mu\nu}$
| then recall identity $\frac{\partial \mathcal{L}}{\partial(\partial_\mu A_\sigma)} = -\frac{1}{\mu_0}F^{\mu\sigma}$

Solutions to Exercises for Part III

$$\mu_0 \Theta^{\mu\nu} = -F^{\mu\sigma}(\partial^\nu A_\sigma) + \frac{1}{4}\eta^{\mu\nu} F^{\alpha\beta} F_{\alpha\beta} \tag{14.14}$$

| perform gauge transformation with $A'_\sigma = A_\sigma + \partial_\sigma \lambda$ and $F'^{\mu\sigma} = F^{\mu\sigma}$

$$\mu_0 \Theta'^{\mu\nu} = -F^{\mu\sigma}(\partial^\nu A_\sigma + \partial^\nu \partial_\sigma \lambda) + \frac{1}{4}\eta^{\mu\nu} F^{\alpha\beta} F_{\alpha\beta} \tag{14.15}$$

$$= \mu_0 \Theta^{\mu\nu} - F^{\mu\sigma} \partial^\nu \partial_\sigma \lambda \tag{14.16}$$

We recall $\chi^{\alpha\mu\nu} = -\frac{1}{\mu_0} F^{\alpha\mu} A^\nu$ and so we take the 4-divergence,

$$\mu_0 \partial_\sigma \chi^{\sigma\mu\nu} = -\partial_\sigma(F^{\sigma\mu} A^\nu) \tag{14.17}$$

| undergo a gauge transformation

$$\mu_0 \partial_\sigma \chi'^{\sigma\mu\nu} = -\partial_\sigma(F^{\sigma\mu}(A^\nu + \partial^\nu \lambda)) \tag{14.18}$$

$$= \mu_0 \partial_\sigma \chi^{\sigma\mu\nu} - \partial_\sigma(F^{\sigma\mu} \partial^\nu \lambda) \tag{14.19}$$

Finally, we can check the gauge transformation of the symmetric energy-momentum tensor.

$$T'^{\mu\nu} = \Theta'^{\mu\nu} + \partial_\sigma \chi'^{\sigma\mu\nu} \tag{14.20}$$

$$= \Theta^{\mu\nu} - F^{\mu\sigma} \partial^\nu \partial_\sigma \lambda + \partial_\sigma \chi^{\sigma\mu\nu} - \partial_\sigma(F^{\sigma\mu} \partial^\nu \lambda) \tag{14.21}$$

$$= T^{\mu\nu} - F^{\mu\sigma} \partial^\nu \partial_\sigma \lambda - (\partial_\sigma F^{\sigma\mu})(\partial^\nu \lambda) - F^{\sigma\mu} \partial_\sigma \partial^\nu \lambda \tag{14.22}$$

| the 3rd term is zero because of source free Maxwell equation $\partial_\sigma F^{\sigma\mu} = 0$
| 2nd and 4th term cancels because $F^{\sigma\mu} = -F^{\mu\sigma}$

$$= T^{\mu\nu} \tag{14.23}$$

Hence indeed adding the extra term $\partial_\sigma \chi^{\sigma\mu\nu}$ makes the energy-momentum tensor symmetric and gauge invariant!

5. Assume it is an Action and we write it out:

$$S = \int L dt = \int \int F_{\mu\nu} \tilde{F}^{\mu\nu} d^3\vec{x} dt \tag{14.24}$$

$$\propto \int \int \vec{E} \cdot \vec{B} \, d^3\vec{x} dt \tag{14.25}$$

| use the expressions for $\vec{E}$ and $\vec{B}$

$$= \int \int \left(-\vec{\nabla}\phi - \frac{\partial \vec{A}}{\partial t}\right) \cdot \left(\vec{\nabla} \times \vec{A}\right) d^3\vec{x} dt \tag{14.26}$$

| use vector identity: $\vec{\nabla} \cdot (\vec{A} \times \vec{B}) = \vec{B} \cdot (\vec{\nabla} \times \vec{A}) - \vec{A} \cdot (\vec{\nabla} \times \vec{B})$
| note that $\frac{\partial \vec{A}}{\partial t} \cdot (\vec{\nabla} \times \vec{A}) = 0$ and $\vec{\nabla} \times \vec{\nabla}\phi = 0$
| so $-\vec{\nabla}\phi \cdot (\vec{\nabla} \times \vec{A}) = \vec{\nabla} \cdot (\vec{\nabla}\phi \times \vec{A})$

$$= \int \left(\int \vec{\nabla} \cdot (\vec{\nabla}\phi \times \vec{A}) \, d^3\vec{x}\right) dt \tag{14.27}$$

which is a surface term after using divergence theorem. Hence this term is not suitable as surface terms are typically assumed to fall off to zero towards infinity and hence the "flux" integral would be zero in most cases.

Part V

Appendices

15
Mathematics: Vector Calculus

15.1 3 Coordinate systems

15.1.1 Cartesian coordinates

Elements:

$\vec{dl} = dx\hat{x} + dy\hat{y} + dz\hat{z}$ and $dV = dxdydz$

Gradient:

$$\vec{\nabla} f = \frac{\partial f}{\partial x}\hat{x} + \frac{\partial f}{\partial y}\hat{y} + \frac{\partial f}{\partial z}\hat{z}$$

Divergence:

$$\vec{\nabla} \cdot \vec{A} = \frac{\partial A_x}{\partial x} + \frac{\partial A_y}{\partial y} + \frac{\partial A_z}{\partial z}$$

Curl:

$$\vec{\nabla} \times \vec{A} = \left(\frac{\partial A_z}{\partial y} - \frac{\partial A_y}{\partial z}\right)\hat{x} + \left(\frac{\partial A_x}{\partial z} - \frac{\partial A_z}{\partial x}\right)\hat{y} + \left(\frac{\partial A_y}{\partial x} - \frac{\partial A_x}{\partial y}\right)\hat{z}$$

Laplacian:

$$\vec{\nabla}^2 f = \frac{\partial^2 f}{\partial x^2} + \frac{\partial^2 f}{\partial y^2} + \frac{\partial^2 f}{\partial z^2}$$

15.1.2 Spherical coordinates

Elements:

$\vec{dl} = dr\hat{r} + rd\theta\hat{\theta} + r\sin\theta d\phi\hat{\phi}$ and $dV = r^2 \sin\theta dr d\theta d\phi$

Gradient:

$$\vec{\nabla} f = \frac{\partial f}{\partial r}\hat{r} + \frac{1}{r}\frac{\partial f}{\partial \theta}\hat{\theta} + \frac{1}{r\sin\theta}\frac{\partial f}{\partial \phi}\hat{\phi}$$

Divergence:

$$\vec{\nabla} \cdot \vec{A} = \frac{1}{r^2}\frac{\partial}{\partial r}(r^2 A_r) + \frac{1}{r\sin\theta}\frac{\partial}{\partial \theta}(A_\theta \sin\theta) + \frac{1}{r\sin\theta}\frac{\partial A_\phi}{\partial \phi}$$

Curl:

$$\vec{\nabla} \times \vec{A} = \frac{1}{r\sin\theta}\left(\frac{\partial}{\partial\theta}(A_\phi \sin\theta) - \frac{\partial A_\theta}{\partial\phi}\right)\hat{r} + \frac{1}{r}\left(\frac{1}{\sin\theta}\frac{\partial A_r}{\partial\phi} - \frac{\partial}{\partial r}(rA_\phi)\right)\hat{\theta} + \frac{1}{r}\left(\frac{\partial}{\partial r}(rA_\theta) - \frac{\partial A_r}{\partial\theta}\right)\hat{\phi}$$

Laplacian:

$$\vec{\nabla}^2 f = \frac{1}{r^2}\frac{\partial}{\partial r}\left(r^2\frac{\partial f}{\partial r}\right) + \frac{1}{r^2\sin\theta}\frac{\partial}{\partial\theta}\left(\sin\theta\frac{\partial f}{\partial\theta}\right) + \frac{1}{r^2\sin^2\theta}\frac{\partial^2 f}{\partial\phi^2}$$

15.1.3 Cylindrical coordinates

Elements:

$d\vec{l} = ds\,\hat{s} + s\,d\phi\,\hat{\phi} + dz\,\hat{z}$ and $dV = s\,ds\,d\phi\,dz$

Gradient:

$$\vec{\nabla} f = \frac{\partial f}{\partial s}\hat{s} + \frac{1}{s}\frac{\partial f}{\partial\phi}\hat{\phi} + \frac{\partial f}{\partial z}\hat{z}$$

Divergence:

$$\vec{\nabla}\cdot\vec{A} = \frac{1}{s}\frac{\partial}{\partial s}(sA_s) + \frac{1}{s}\frac{\partial A_\phi}{\partial\phi} + \frac{\partial A_z}{\partial z}$$

Curl:

$$\vec{\nabla}\times\vec{A} = \left(\frac{1}{s}\frac{\partial A_z}{\partial\phi} - \frac{\partial A_\phi}{\partial z}\right)\hat{s} + \left(\frac{\partial A_s}{\partial z} - \frac{\partial A_z}{\partial s}\right)\hat{\phi} + \frac{1}{s}\left(\frac{\partial}{\partial s}(sA_\phi) - \frac{\partial A_s}{\partial\phi}\right)\hat{z}$$

Laplacian:

$$\vec{\nabla}^2 f = \frac{1}{s}\frac{\partial}{\partial s}\left(s\frac{\partial f}{\partial s}\right) + \frac{1}{s^2}\frac{\partial^2 f}{\partial\phi^2} + \frac{\partial^2 f}{\partial z^2}$$

15.2 Vector identities

15.2.1 Triple products

1. $\vec{A}\cdot(\vec{B}\times\vec{C}) = \vec{B}\cdot(\vec{C}\times\vec{A}) = \vec{C}\cdot(\vec{A}\times\vec{B})$
2. $\vec{A}\times(\vec{B}\times\vec{C}) = \vec{B}(\vec{A}\cdot\vec{C}) - \vec{C}(\vec{A}\cdot\vec{B})$

15.2.2 Product rules

1. $\vec{\nabla}(fg) = f(\vec{\nabla}g) + g(\vec{\nabla}f)$
2. $\vec{\nabla}(\vec{A}\cdot\vec{B}) = \vec{A}\times(\vec{\nabla}\times\vec{B}) + \vec{B}\times(\vec{\nabla}\times\vec{A}) + (\vec{A}\cdot\vec{\nabla})\vec{B} + (\vec{B}\cdot\vec{\nabla})\vec{A}$
3. $\vec{\nabla}\cdot(f\vec{A}) = f(\vec{\nabla}\cdot\vec{A}) + \vec{A}\cdot(\vec{\nabla}f)$
4. $\vec{\nabla}\cdot(\vec{A}\times\vec{B}) = \vec{B}\cdot(\vec{\nabla}\times\vec{A}) - \vec{A}\cdot(\vec{\nabla}\times\vec{B})$

Mathematics: Vector Calculus 145

5. $\vec{\nabla} \times (f\vec{A}) = f(\vec{\nabla} \times \vec{A}) - \vec{A} \times (\vec{\nabla} f)$
6. $\vec{\nabla} \times (\vec{A} \times \vec{B}) = (\vec{B} \cdot \vec{\nabla})\vec{A} - (\vec{A} \cdot \vec{\nabla})\vec{B} + \vec{A}(\vec{\nabla} \cdot \vec{B}) - \vec{B}(\vec{\nabla} \cdot \vec{A})$

15.2.3 Second derivatives

1. $\vec{\nabla} \cdot (\vec{\nabla} \times \vec{A}) = 0$
2. $\vec{\nabla} \times (\vec{\nabla} f) = 0$
3. $\vec{\nabla} \times (\vec{\nabla} \times \vec{A}) = \vec{\nabla}(\vec{\nabla} \cdot \vec{A}) - \vec{\nabla}^2 \vec{A}$

15.2.4 Integral theorems

1. Gradient Theorem: $\int_a^b (\vec{\nabla} f) \cdot d\vec{l} = f(b) - f(a)$

2. Divergence or, Gauss or, Green's Theorem: $\int (\vec{\nabla} \cdot \vec{A}) dV = \oint \vec{A} \cdot d\vec{A}$

3. Stokes Theorem: $\int (\vec{\nabla} \times \vec{A}) \cdot d\vec{A} = \oint \vec{A} \cdot d\vec{l}$

15.2.5 Dirac delta function

- Definition: $\vec{\nabla} \cdot \left(\dfrac{\hat{r}}{r^2} \right) = 4\pi \delta^3(\vec{r})$

- A useful related expression is $\vec{\nabla}^2 \dfrac{1}{r} = \vec{\nabla} \cdot \left(\vec{\nabla} \dfrac{1}{r} \right) = \vec{\nabla} \cdot \left(-\dfrac{\hat{r}}{r^2} \right) = -4\pi \delta^3(\vec{r})$

16

Summary of Electromagnetism

This appendix is a summary of electrodynamics at the junior undergraduate level - which is usually named "Electromagnetism". Electromagnetism is mostly about electrostatics and magnetostatics. Even though Faraday's law is about "dynamics" and has no "static" version, it has to be included since it is one the four Maxwell's equations.

This appendix includes some exercises (with solutions) for the student to recap on the basic applications in Electromagnetism.

16.1 Electrostatics

- Coulomb's Law: $\vec{F} = \dfrac{1}{4\pi\epsilon_0} \dfrac{qQ}{|\vec{r}-\vec{r}'|^2} \dfrac{\vec{r}-\vec{r}'}{|\vec{r}-\vec{r}'|} = \dfrac{1}{4\pi\epsilon_0} \dfrac{qQ}{|\vec{r}-\vec{r}'|^3}(\vec{r}-\vec{r}')$ where $\vec{r}$ is the position vector of charge Q, $\vec{r}'$ is the position vector of the charge q. We use a new notation to simplify writing of $\vec{r}-\vec{r}'$, let $\vec{R} = \vec{r}-\vec{r}'$, so $R = |\vec{r}-\vec{r}'|$ and $\hat{R} = \dfrac{\vec{r}-\vec{r}'}{|\vec{r}-\vec{r}'|}$. Then $\vec{F} = \dfrac{1}{4\pi\epsilon_0} \dfrac{qQ}{R^2} \hat{R}$.

- Electric Force: $\vec{F} = Q\vec{E}$

- Discrete charges (Superposition principle): $\vec{E}(\vec{r}) = \sum_{i=1}^{n} \vec{E}_i = \dfrac{1}{4\pi\epsilon_0} \sum_{i=1}^{n} \dfrac{q_i}{|\vec{r}-\vec{r}_i|^2} \hat{R}_i$

- Continuous charges (Superposition principle): $\vec{E}(\vec{r}) = \dfrac{1}{4\pi\epsilon_0} \int \dfrac{\rho(\vec{r}')}{|\vec{r}-\vec{r}'|^2} \hat{R} \, dV'$

- Gauss Law: $\oint \vec{E} \cdot d\vec{A} = \dfrac{Q_{\text{enc}}}{\epsilon_0} \xrightarrow{\text{divergence theorem}} \vec{\nabla} \cdot \vec{E} = \dfrac{\rho}{\epsilon_0}$

- Curl[1] of $\vec{E}$: $\oint \vec{E} \cdot d\vec{l} = 0 \xrightarrow{\text{Stokes theorem}} \vec{\nabla} \times \vec{E} = 0$

- Scalar Potential[2]: $\oint \vec{E} \cdot d\vec{l} = 0 \xrightarrow{\text{can define}} \phi(\vec{r}_2) - \phi(\vec{r}_1) = -\int_{\vec{r}_1}^{\vec{r}_2} \vec{E} \cdot d\vec{l} \xrightarrow{\text{Gradient theorem}}$
  $\vec{E} = -\vec{\nabla}\phi$. Or we can proceed from $\vec{\nabla} \times \vec{E} = 0 \xrightarrow{\vec{\nabla}\times(\vec{\nabla}f)=0} \vec{E} = -\vec{\nabla}\phi$

- Poisson's equation: $\vec{\nabla} \cdot \vec{E} = \dfrac{\rho}{\epsilon_0} \xrightarrow{\vec{E}=-\vec{\nabla}\phi} \vec{\nabla} \cdot (-\vec{\nabla}\phi) = \dfrac{\rho}{\epsilon_0}$ so $\vec{\nabla}^2\phi = -\dfrac{\rho}{\epsilon_0}$. The solution for PE $=0$ at ∞ is $\phi(\vec{r}) = \dfrac{1}{4\pi\epsilon_0} \int \dfrac{\rho(\vec{r}')}{|\vec{r}-\vec{r}'|} dV'$

---

[1] In electrostatics, $\vec{E}$ is a conservative field but not in electrodynamics.
[2] The negative sign is largely a matter of convention.

Summary of Electromagnetism

- Laplace's equation: $\vec{\nabla}^2 \phi = 0$ for regions with no charge.
  - Uniqueness Theorem 1: The solution to Laplace's equation in some volume V is uniquely determined if ϕ is specified on the boundary surface. This theorem gives rise to the "method of images" trick.
  - Multipole expansion:

$$\phi(\vec{r}) = \frac{1}{4\pi\epsilon_0} \int \frac{\rho(\vec{r}')}{|\vec{r}-\vec{r}'|} dV'$$

$$= \frac{1}{4\pi\epsilon_0} \sum_{n=0}^{\infty} \frac{1}{r^{n+1}} \int r'^n P_n(\cos\theta') \rho(\vec{r}') dV'$$

  where P_n is the nth Legendre Polynomial

$$= \frac{1}{4\pi\epsilon_0} \left[\frac{1}{r} \int \rho(\vec{r}') dV' + \frac{1}{r^2} \int r' \cos\theta' \rho(\vec{r}') dV' + \frac{1}{r^3} \int r'^2 \left(\frac{3}{2}\cos^2\theta' - \frac{1}{2} \right) \rho(\vec{r}') dV' + \cdots \right]$$

  * Dipole case: Potential : $\phi(\vec{r}) = \frac{1}{4\pi\epsilon_0} \frac{\vec{p} \cdot \hat{r}}{r^2}$ where $\vec{p}$ is the dipole moment.
    Electric field: $\vec{E}(r, \theta) = \frac{p}{4\pi\epsilon_0 r^3}\left(2\cos\theta \hat{r} + \sin\theta \hat{\theta} \right)$

- Boundary conditions
  1. $E^{\perp}_{\text{above}} - E^{\perp}_{\text{below}} = \frac{\sigma}{\epsilon_0}$ where σ is the surface charge
  2. $\vec{E}^{\parallel}_{\text{above}} = \vec{E}^{\parallel}_{\text{below}}$
  3. $\phi_{\text{above}} = \phi_{\text{below}}$

- Energy stored: $W = \frac{\epsilon_0}{2} \int_{\text{all space}} \vec{E}^2 dV$

16.2 Electric fields in matter

- Induced dipole moment: $\vec{p} = \overleftrightarrow{\alpha} \vec{E}$ for Linear materials and $\overleftrightarrow{\alpha}$ is the polarisability 2-tensor,
  $\overleftrightarrow{\alpha} = \begin{pmatrix} \alpha_{xx} & \alpha_{xy} & \alpha_{xz} \\ \alpha_{yx} & \alpha_{yy} & \alpha_{yz} \\ \alpha_{zx} & \alpha_{zy} & \alpha_{zz} \end{pmatrix}$. If the material is also isotropic (LI), then $\overleftrightarrow{\alpha} = \alpha \mathcal{I}_{3\times 3}$ where $\mathcal{I}_{3\times 3}$ is the 3-dimensional identity matrix and α is the polarisability constant.

- Force on electric dipole: $\vec{F} = (\vec{p} \cdot \vec{\nabla}) \vec{E}$

- Torque on electric dipole: $\vec{\tau} = (\vec{p} \times \vec{E}) + (\vec{r} \times \vec{F})$

- Polarisation: $\vec{P} = $ dipole moment per unit volume.

- Potential of polarised object:

$$\phi(\vec{r}) = \frac{1}{4\pi\epsilon_0} \int \frac{\hat{R} \cdot \vec{P}(\vec{r}')}{|\vec{r} - \vec{r}'|^2} dV' \tag{16.1}$$

$$= \frac{1}{4\pi\epsilon_0} \oint \frac{1}{|\vec{r} - \vec{r}'|} \vec{P} \cdot d\vec{A}' - \frac{1}{4\pi\epsilon_0} \int \frac{1}{|\vec{r} - \vec{r}'|} \left(\vec{\nabla}' \cdot \vec{P}\right) dV' \tag{16.2}$$

$$= \frac{1}{4\pi\epsilon_0} \oint \frac{\sigma_b}{|\vec{r} - \vec{r}'|} dA' + \frac{1}{4\pi\epsilon_0} \int \frac{\rho_b}{|\vec{r} - \vec{r}'|} dV' \tag{16.3}$$

where the bound surface charge is defined as $\sigma_b = \vec{P} \cdot \hat{n}$ (and $d\vec{A}' = \hat{n} dA'$), and the bound volume charge is defined as $\rho_b = -\vec{\nabla} \cdot \vec{P}$.

- Displacement field $\vec{D} = \epsilon_0 \vec{E} + \vec{P}$: From Gauss law, $\epsilon_0 \vec{\nabla} \cdot \vec{E} = \rho_b + \rho_f = -\vec{\nabla} \cdot \vec{P} + \rho_f \implies \vec{\nabla} \cdot (\epsilon_0 \vec{E} + \vec{P}) = \rho_f \implies \vec{\nabla} \cdot \vec{D} = \rho_f$. Note that $\vec{\nabla} \times \vec{D} = \vec{\nabla} \times \vec{P} \neq 0$.

- Boundary conditions: $D^\perp_{\text{above}} - D^\perp_{\text{below}} = \sigma_f$ and $\vec{D}^\parallel_{\text{above}} - \vec{D}^\parallel_{\text{below}} = \vec{P}^\parallel_{\text{above}} - \vec{P}^\parallel_{\text{below}}$.

- (Dimensionless) Electric susceptibility $\overleftrightarrow{\chi}_e$: the so-called constitutive relation is $\vec{P} = \epsilon_0 \overleftrightarrow{\chi}_e \vec{E}$ and $\overleftrightarrow{\chi}_e$ is the electric susceptibility 2-tensor. In LI materials, $\vec{P} = \epsilon_0 \chi_e \vec{E}$.

- Permittivity $\overleftrightarrow{\epsilon} = \mathcal{I} + \overleftrightarrow{\chi}_e$: For Linear media, $\vec{D} = \epsilon_0 \vec{E} + \vec{P} = \epsilon_0 \vec{E} + \epsilon_0 \overleftrightarrow{\chi}_e \vec{E} = \overleftrightarrow{\epsilon} \vec{E}$. For LI materials, permittivity is a constant $\epsilon = 1 + \chi_e$.

- Relative permittivity or dielectric constant $\overleftrightarrow{\epsilon}_r = \frac{1}{\epsilon_0} \overleftrightarrow{\epsilon} = \frac{1}{\epsilon_0} \left(\mathcal{I} + \overleftrightarrow{\chi}_e\right)$ and again in LI materials, they are all constants.

- Energy in Dielectric systems: $W = \frac{1}{2} \int \vec{D} \cdot \vec{E} dV$ and in vacuum, $\chi_e = 0$ and so $\vec{D} = \epsilon_0 \vec{E}$ gives back the vacuum result: $W = \frac{\epsilon_0}{2} \int \vec{E}^2 dV$.

16.3 Magnetostatics

- Magnetic force: $\vec{F} = Q(\vec{v} \times \vec{B})$, $\vec{F} = \int I(d\vec{l} \times \vec{B})$ and $\vec{F} = \int \vec{J} \times \vec{B} dV$ where $\vec{J}$ is the volume current density.

- Lorentz force law: $\vec{F} = Q(\vec{E} + \vec{v} \times \vec{B})$

- Biot-Savart law: Magnetic field of a steady current is $\vec{B}(\vec{r}) = \frac{\mu_0 I}{4\pi} \int \frac{d\vec{l} \times \hat{R}}{R^2}$. For surface current density $\vec{K}$, $\vec{B}(\vec{r}) = \frac{\mu_0}{4\pi} \int \frac{\vec{K}(\vec{r}') \times \hat{R}}{R^2} dA'$. For volume current density $\vec{J}$, $\vec{B}(\vec{r}) = \frac{\mu_0}{4\pi} \int \frac{\vec{J}(\vec{r}') \times \hat{R}}{R^2} dV'$. For point charge, $\vec{B}(\vec{r}) = \frac{\mu_0}{4\pi} \frac{q\vec{v} \times \hat{R}}{R^2}$ but it only an approximation because Biot-Savart law is for steady currents and a moving charge is not, also we require $v \ll c$ as we will see in Part 1.

- Divergence of $\vec{B}$: $\oint \vec{B} \cdot d\vec{A} = 0 \xrightarrow{\text{Divergence theorem}} \vec{\nabla} \cdot \vec{B} = 0$.

Summary of Electromagnetism

- Curl of $\vec{B}$: Use Ampere's law, $\oint \vec{B} \cdot d\vec{l} = \mu_0 I_{enc} \xrightarrow{\text{Stokes theorem}} \vec{\nabla} \times \vec{B} = \mu_0 \vec{J}$.

- Magnetic Vector Potential: Using $\vec{\nabla} \cdot \vec{B} = 0 \xrightarrow{\vec{\nabla} \cdot (\vec{\nabla} \times \vec{A})=0} \vec{B} = \vec{\nabla} \times \vec{A}$ and the vector potential, $\vec{A}$ has freedom: $\vec{A}' = \vec{A} + \vec{\nabla}\lambda$ since both vector potentials give the same physical magnetic field: $\vec{\nabla} \times \vec{A}' = \vec{\nabla} \times \vec{A} = \vec{B}$. We can carry this freedom over to $\vec{\nabla} \cdot \vec{A}'$. Setting $\vec{\nabla} \cdot \vec{A}' = 0$ is called the Coulomb gauge.

- Boundary conditions:

  1. $B_{above}^{\perp} = B_{below}^{\perp}$
  2. $B_{above}^{\parallel} - B_{below}^{\parallel} = \mu_0 K$ where K is the surface current
  3. For Coulomb gauge $\vec{\nabla} \cdot \vec{A} = 0$, we have $\vec{A}_{above} = \vec{A}_{below}$ and $\dfrac{\partial \vec{A}_{above}}{\partial n} - \dfrac{\partial \vec{A}_{below}}{\partial n} = -\mu_0 \vec{K}$.

- Magnetic Dipole: Although a multipole expansion of $\vec{A}$ can be done, the dipole contribution is a good approximation for the far field situation. The dipole vector potential is $\vec{A}(\vec{r}) = \dfrac{\mu_0}{4\pi} \dfrac{\vec{m} \times \hat{r}}{r^2}$ where $\vec{m} = I \int d\vec{A} = I\vec{A}$ is the magnetic dipole moment.

16.4 Magnetic fields in matter

- Torque on magnetic dipole: $\vec{\tau} = \vec{m} \times \vec{B}$

- Force on magnetic dipole: $\vec{F} = \vec{\nabla}(\vec{m} \cdot \vec{B})$

- Magnetisation, $\vec{M}$, is defined to be the magnetic dipole moment per unit volume.

- Vector potential of a magnetised object: $\vec{A}(\vec{r}) = \dfrac{\mu_0}{4\pi} \int \dfrac{\vec{M}(\vec{r}') \times \hat{R}}{R^2} dV' = \dfrac{\mu_0}{4\pi} \int \dfrac{\vec{J}_b(\vec{r}')}{R} dV' + \dfrac{\mu_0}{4\pi} \oint \dfrac{\vec{K}_b(\vec{r}')}{R} dA'$ where the bound volume current is $\vec{J}_b = \vec{\nabla} \times \vec{M}$ and the bound surface current is $\vec{K}_b = \vec{M} \times \hat{n}$.

- Ampere's law in magnetised materials:

$$\frac{1}{\mu_0}(\vec{\nabla} \times \vec{B}) = \vec{J} = \vec{J}_f + \vec{J}_b = \vec{J}_f + (\vec{\nabla} \times \vec{M})$$

$$\Longrightarrow \vec{\nabla} \times \left(\frac{1}{\mu_0}\vec{B} - \vec{M}\right) = \vec{J}_f$$

$$\vec{\nabla} \times \vec{H} = \vec{J}_f \xrightarrow{\text{in integral form}} \oint \vec{H} \cdot d\vec{l} = I_{f,enc}$$

- Divergence: $\vec{\nabla} \cdot \vec{B} = 0$ but note that $\vec{H}$ does not have zero divergence since $\vec{\nabla} \cdot \vec{H} = -\vec{\nabla} \cdot \vec{M} \neq 0$.

- Boundary conditions: $H_{above}^{\perp} - H_{below}^{\perp} = -(M_{above}^{\perp} - M_{below}^{\perp})$ and $\vec{H}_{above}^{\parallel} - \vec{H}_{below}^{\parallel} = \vec{K}_f \times \hat{n}$.

- Magnetic susceptibility: For linear media, $\vec{M} = \overset{\leftrightarrow}{\chi}_m \vec{H}$ and in LI materials, χ_m is a number. For diamagnets, $\chi_m < 0$ and for paramagnets, $\chi_m > 0$.

- Permeability: $\vec{B} = \mu_0(\vec{H} + \vec{M}) = \mu_0(1 + \overset{\leftrightarrow}{\chi}_m)\vec{M} = \overset{\leftrightarrow}{\mu}\vec{H}$ where $\overset{\leftrightarrow}{\mu}$ is the permeability. For vacuum, $\mu = \mu_0$ as $\chi_m = 0$.

- Note: For LI materials, we have $\vec{J}_b = \vec{\nabla} \times \vec{M} = \vec{\nabla} \times (\chi_m \vec{H}) = \chi_m \vec{\nabla} \times \vec{H} = \chi_m \vec{J}_f$. Thus unless free current actually flows through the material, all bound current will be at the surface.

16.5 Further topics

- Ohm's law: $\vec{J} = \overset{\leftrightarrow}{\sigma} \vec{E}$ where $\overset{\leftrightarrow}{\sigma}$ is the electrical conductivity tensor. For LI materials, $\overset{\leftrightarrow}{\sigma} = \sigma \mathcal{I}_{3 \times 3}$. The other version of Ohm's law is $V = IR$ where V is the (electric) potential difference and R is the resistance.

- Joule heating law: $P = VI = I^2 R$ where P is the power delivered to the resistor.

- Faraday's law: $\oint \vec{E} \cdot d\vec{l} = -\dfrac{d\Phi_B}{dt}$ where Φ_B is the magnetic flux and $\vec{E}$ is the "induced" electric field. Then we convert it into the differential version: $\oint \vec{E} \cdot d\vec{l} = -\dfrac{d\Phi_B}{dt} = -\int \dfrac{\partial \vec{B}}{\partial t} \cdot d\vec{A} \overset{\text{Stokes theorem}}{\longrightarrow} \vec{\nabla} \times \vec{E} = -\dfrac{\partial \vec{B}}{\partial t}$. Note that motional EMF (or "flux cutting") is slightly different; it is caused by a magnetic force.

- Energy in magnetic fields: $W = \dfrac{1}{2\mu_0} \int_{\text{all space}} \vec{B}^2 \, dV$

- Maxwell's fix for Ampere's law (displacement current), take divergence of Ampere's law:

$$\begin{aligned}
\vec{\nabla} \cdot (\vec{\nabla} \times \vec{B}) &= \mu_0 \vec{\nabla} \cdot \vec{J} \\
&\quad | \text{ note that the left hand side is mathematically zero} \\
&\quad | \text{ use continuity equation on the right hand side} \\
0 &= -\mu_0 \dfrac{\partial \rho}{\partial t} \\
&\quad | \text{ use Gauss law} \\
0 &= -\mu_0 \dfrac{\partial}{\partial t}(\epsilon_0 \vec{\nabla} \cdot \vec{E}) \\
0 &= -\mu_0 \epsilon_0 \vec{\nabla} \cdot \dfrac{\partial \vec{E}}{\partial t}
\end{aligned}$$

Thus we arrive at a contradictory situation where the left side is zero but the right side may not be. The corrected Ampere's law which has zero divergence on the right side, is $\vec{\nabla} \times \vec{B} = \mu_0 \vec{J} + \mu_0 \epsilon_0 \dfrac{\partial \vec{E}}{\partial t}$. In other words, we can say that the corrected Ampere's law is now compatible with the charge continuity equation.

Summary of Electromagnetism 151

- Maxwell's equations:

    1. Gauss law: $\vec{\nabla} \cdot \vec{E} = \dfrac{\rho}{\epsilon_0}$
    2. No monopoles: $\vec{\nabla} \cdot \vec{B} = 0$
    3. Faraday's law: $\vec{\nabla} \times \vec{E} = -\dfrac{\partial \vec{B}}{\partial t}$
    4. Ampere–Maxwell's law: $\vec{\nabla} \times \vec{B} = \mu_0 \vec{J} + \mu_0 \epsilon_0 \dfrac{\partial \vec{E}}{\partial t}$
    5. (Supplemented by) Lorentz force law: $\vec{F} = q(\vec{E} + \vec{v} \times \vec{B})$

- Maxwell's equations rewritten to be more suitable for materials.

    1. $\vec{\nabla} \cdot \vec{D} = \rho_f$
    2. $\vec{\nabla} \cdot \vec{B} = 0$
    3. $\vec{\nabla} \times \vec{E} = -\dfrac{\partial \vec{B}}{\partial t}$
    4. $\vec{\nabla} \times \vec{H} = \vec{J}_f + \dfrac{\partial \vec{D}}{\partial t}$
    5. Constitutive relations: In the case of LI materials, $\vec{P} = \epsilon_0 \chi_e \vec{E} \Rightarrow \vec{D} = \epsilon \vec{E}$ and $\vec{M} = \chi_m \vec{H} \Rightarrow \vec{H} = \dfrac{1}{\mu} \vec{B}$ where $\epsilon = \epsilon_0(1 + \chi_e)$ and $\mu = \mu_0(1 + \chi_m)$.

- Boundary conditions (medium 1 and medium 2):

    1. $D_1^\perp - D_2^\perp = \sigma_f$
    2. $B_1^\perp - B_2^\perp = 0$
    3. $\vec{E}_1^{\|} - \vec{E}_2^{\|} = 0$
    4. $\vec{H}_1^{\|} - \vec{H}_2^{\|} = \vec{K}_f \times \hat{n}$

For the case where there is no free charge or free current at the interface,

1. $\epsilon_1 E_1^\perp - \epsilon_2 E_2^\perp = 0$
2. $B_1^\perp - B_2^\perp = 0$
3. $\vec{E}_1^{\|} - \vec{E}_2^{\|} = 0$
4. $\dfrac{1}{\mu_1} \vec{B}_1^{\|} - \dfrac{1}{\mu_2} \vec{B}_2^{\|} = 0$

16.6 Electromagnetic waves (Standard application of electrodynamics)

- Wave equation for $\vec{E}$ and $\vec{B}$:

    - For no charge or current, Maxwell's equations are: $\vec{\nabla} \cdot \vec{E} = 0$, $\vec{\nabla} \cdot \vec{B} = 0$, $\vec{\nabla} \times \vec{E} = -\dfrac{\partial \vec{B}}{\partial t}$ and $\vec{\nabla} \times \vec{B} = \mu_0 \epsilon_0 \dfrac{\partial \vec{E}}{\partial t}$

- Take curl of Faraday's law:

$$\vec{\nabla} \times (\vec{\nabla} \times \vec{E}) = \vec{\nabla} \times \left(-\frac{\partial \vec{B}}{\partial t}\right)$$

$$\Rightarrow \vec{\nabla}(\vec{\nabla} \cdot \vec{E}) - \vec{\nabla}^2 \vec{E} = -\frac{\partial}{\partial t}(\vec{\nabla} \times \vec{B})$$

$$\Rightarrow \vec{\nabla}^2 \vec{E} = \mu_0 \epsilon_0 \frac{\partial^2 \vec{E}}{\partial t^2}$$

- Take curl of Ampere–Maxwell's law:

$$\vec{\nabla} \times (\vec{\nabla} \times \vec{B}) = \vec{\nabla} \times \left(\mu_0 \epsilon_0 \frac{\partial \vec{E}}{\partial t}\right)$$

$$\Rightarrow \vec{\nabla}(\vec{\nabla} \cdot \vec{B}) - \vec{\nabla}^2 \vec{B} = \mu_0 \epsilon_0 \frac{\partial}{\partial t}(\vec{\nabla} \times \vec{E})$$

$$\Rightarrow \vec{\nabla}^2 \vec{B} = \mu_0 \epsilon_0 \frac{\partial^2 \vec{B}}{\partial t^2}$$

- These are wave equations for waves traveling at a speed $v = \frac{1}{\sqrt{\mu_0 \epsilon_0}} = c$ and the general solution is of the form: $\vec{B}(z,t), \vec{E}(z,t) = \vec{f}(z - ct) + \vec{g}(z + ct)$ where $\vec{f}$ is the wave moving towards increasing z and $\vec{g}$ is the wave moving towards decreasing z.

- EM waves are transverse due to $\vec{\nabla} \cdot \vec{E} = 0$ and $\vec{\nabla} \cdot \vec{B} = 0$.

- $\vec{E}$ and $\vec{B}$ are in phase and are mutually perpendicular due to Faraday's law: $\vec{\nabla} \times \vec{E} = -\frac{\partial \vec{B}}{\partial t}$.

- Three laws of geometrical optics can be derived.

    1. Incident, reflected and transmitted wave vectors form a plane.
    2. Angle of incidence = angle of reflection
    3. Snell's law: $n_1 \sin\theta_1 = n_2 \sin\theta_2$

- Brewster's angle: angle where there is no reflected wave but this is for the case where the incident wave has polarisation parallel to the plane of incidence.

16.7 Exercises

16.7.1 Evaluating line integral

For the vector $\vec{v} = y^2 \hat{x} + 2x(y+1)\hat{y}$, evaluate the line integral $\int \vec{v} \cdot d\vec{l}$ from point $a = (1,1)$ to point $b = (2,2)$ via:

1. Path 1: $(1,1) \longrightarrow (2,1) \longrightarrow (2,2)$.
2. Path 2: $(1,1) \longrightarrow (2,2)$ directly.

16.7.2 Proving identity

Prove the identity: $\vec{\nabla} \cdot (\vec{A} \times \vec{B}) = \vec{B} \cdot (\vec{\nabla} \times \vec{A}) - \vec{A} \cdot (\vec{\nabla} \times \vec{B})$.

16.7.3 Proving identity

Prove the identity: $\vec{\nabla} \times (\vec{\nabla} \times \vec{A}) = \vec{\nabla}(\vec{\nabla} \cdot \vec{A}) - \vec{\nabla}^2 \vec{A}$

16.7.4 Electrostatics

For the electrostatics scalar potential $\phi(\vec{r}) = \dfrac{1}{4\pi\epsilon_0} \int \dfrac{\rho(\vec{r}')}{|\vec{r} - \vec{r}'|} dV'$, work out explicitly to show that the equation $\vec{\nabla} \times \vec{E} = 0$ is satisfied.

16.7.5 Electric dipole

For an electric dipole, the scalar potential is $\phi(\vec{r}) = \dfrac{1}{4\pi\epsilon_0} \dfrac{\vec{p} \cdot \hat{r}}{r^2}$.

1. Now set the dipole moment $\vec{p}$ at the origin and point it in the z-direction, work out explicitly that $\vec{E}(r,\theta) = \dfrac{p}{4\pi\epsilon_0 r^3} \left(2\cos\theta \hat{r} + \sin\theta \hat{\theta} \right)$.

2. Then show that the electric field can be written as $\vec{E} = \dfrac{1}{4\pi\epsilon_0 r^3} (3(\vec{p} \cdot \hat{r})\hat{r} - \vec{p})$.
   (Remark: you may work backwards by starting with $\vec{E} = \dfrac{1}{4\pi\epsilon_0 r^3} (3(\vec{p} \cdot \hat{r})\hat{r} - \vec{p})$ and then impose spherical coordinates.)

16.7.6 Magnetostatics

For the magnetostatics vector potential $\vec{A}(\vec{r}) = \dfrac{\mu_0}{4\pi} \int \dfrac{\vec{J}(\vec{r}')}{|\vec{r} - \vec{r}'|} dV'$, first work out explicitly to show that Coulomb's gauge is satisfied: $\vec{\nabla} \cdot \vec{A} = 0$ and then work out explicitly to show that Ampere's law: $\vec{\nabla} \times \vec{B} = \mu_0 \vec{J}$ is satisfied.

16.7.7 Magnetic field of a surface current

Find the magnetic field of an infinite uniform surface current $\vec{K} = K\hat{x}$, flowing over the xy-plane. (Answer: $\vec{B} = \pm \dfrac{\mu_0}{2} K \hat{y}$ for $z \lessgtr 0$) (Remark: we will need this result in Part II of the main text.)

16.7.8 Poynting vector

Consider a charging circular parallel plate capacitor of radius a and width w with $w \ll a$. A constant current I flows in the charging circuit. You may want to align the circle axis along the z-axis for calculations.

1. Find the electric and magnetic fields in the gap as functions of the distance s from the axis and the time t. (Assume the charge is zero at $t = 0$.)

2. Find the Poynting vector $\vec{S}$ in the gap

3. Calculate the total power entering the gap. (Ignore any fringe effects.)

16.7.9 Electromagnetic waves

1. Verify that this spherical wave (where A is an amplitude factor) $\vec{E} = \dfrac{A}{r}\sin\theta\cos\left(\omega\left(t-\dfrac{r}{c}\right)\right)\hat{\theta}$ satisfies the wave equation $\vec{\nabla}^2\vec{E} = \mu_0\epsilon_0\dfrac{\partial^2\vec{E}}{\partial t^2}$. State and justify any approximations used.

2. In section 16.6, we said that "EM waves are transverse due to $\vec{\nabla}\cdot\vec{E} = 0$". Verify that by using a plane wave solution: $\vec{E} = \vec{E}_0 e^{i(\vec{k}\cdot\vec{r}-\omega t)}$.

16.8 Solutions to Exercises

16.8.1 Solution to exercise 16.7.1

In evaluating the line integral for vector $\vec{v} = y^2\hat{x} + 2x(y+1)\hat{y}$ we always use $\vec{dl} = dx\hat{x} + dy\hat{y}$:

1. For Path 1,

$$\int \vec{v}\cdot\vec{dl} = \int y^2 dx + \int 2x(y+1)dy \tag{16.4}$$

$$= \int_1^2 1^2 dx + \int_1^2 2(2)(y+1)dy \tag{16.5}$$

$$= 1 + 4\int_1^2 (y+1)dy \tag{16.6}$$

$$= 11 \tag{16.7}$$

2. For Path 2, we have the extra condition that $y = x \Longrightarrow dy = dx$

$$\int \vec{v}\cdot\vec{dl} = \int y^2 dx + 2x(y+1)dy \tag{16.8}$$

$$= \int_1^2 y^2 dy + 2y(y+1)dy \tag{16.9}$$

$$= \int_1^2 3y^2 + 2y\, dy \tag{16.10}$$

$$= 10 \tag{16.11}$$

16.8.2 Solution to exercise 16.7.2

We will use the index notation to simplify the calculation.

- Dot product: $\vec{A}\cdot\vec{B} = \sum_i A_i B_i$

- Cross product: $\sum_{ijk} \epsilon_{ijk}\hat{e}_i A_j B_k$

- property 1 of the antisymmetric symbol: (i) for even swaps of indices: $\epsilon_{kij} = \epsilon_{ijk}$, (ii) for odd swaps of indices: $\epsilon_{ikj} = -\epsilon_{ijk}$

Summary of Electromagnetism   155

- property 2 of the antisymmetric symbol: reduction to Kronecker delta: $\sum_k \epsilon_{ijk}\epsilon_{klm} = \delta_{jl}\delta_{km} - \delta_{jm}\delta_{kl}$

$$\text{RHS} = \vec{B}\cdot(\vec{\nabla}\times\vec{A}) - \vec{A}\cdot(\vec{\nabla}\times\vec{B}) \tag{16.12}$$

$$= \sum_{ijk} \epsilon_{ijk} B_i \frac{\partial}{\partial x_j} A_k - \sum_{ijk} A_i \frac{\partial}{\partial x_j} B_k \tag{16.13}$$

| for second term, rename $i \leftrightarrow k$

$$= \sum_{ijk} \left(\epsilon_{ijk} B_i \frac{\partial A_k}{\partial x_j} - \epsilon_{kji} A_k \frac{\partial B_i}{\partial x_j} \right) \tag{16.14}$$

| then $\epsilon_{kji} = -\epsilon_{ijk}$

$$= \sum_{ijk} \epsilon_{ijk} \left(B_i \frac{\partial A_k}{\partial x_j} + A_k \frac{\partial B_i}{\partial x_j} \right) \tag{16.15}$$

$$= \sum_{ijk} \epsilon_{ijk} \frac{\partial}{\partial x_j}(A_k B_i) \tag{16.16}$$

| then $\epsilon_{ijk} = \epsilon_{jki}$

$$= \vec{\nabla}\cdot(\vec{A}\times\vec{B}) \quad \text{hence proved} \tag{16.17}$$

16.8.3 Solution to exercise 16.7.3

Start with the LHS:

$$\text{LHS} = \vec{\nabla}\times(\vec{\nabla}\times\vec{A}) \tag{16.18}$$

$$= \sum_{ijk} \epsilon_{ijk} \hat{e}_i \frac{\partial}{\partial x_j} (\vec{\nabla}\times\vec{A})_k \tag{16.19}$$

$$= \sum_{ijk} \epsilon_{ijk} \hat{e}_i \frac{\partial}{\partial x_j} \sum_{lm} \epsilon_{klm} \frac{\partial}{\partial x_l} A_m \tag{16.20}$$

| use $\sum_k \epsilon_{ijk}\epsilon_{klm} = \sum_k \epsilon_{kij}\epsilon_{klm} = \delta_{il}\delta_{jm} - \delta_{im}\delta_{jl}$

$$= \sum_{ijlm} (\delta_{il}\delta_{jm} - \delta_{im}\delta_{jl})\, \hat{e}_i \frac{\partial^2 A_m}{\partial x_j \partial x_l} \tag{16.21}$$

$$= \sum_{ij} \hat{e}_i \frac{\partial^2 A_j}{\partial x_j \partial x_i} - \sum_{ij} \hat{e}_i \frac{\partial^2 A_i}{\partial x_j^2} \tag{16.22}$$

$$= \sum_{ij} \hat{e}_i \frac{\partial}{\partial x_i}\frac{\partial A_j}{\partial x_j} - \sum_j \frac{\partial \vec{A}}{\partial x_j^2} \tag{16.23}$$

$$= \vec{\nabla}(\vec{\nabla}\cdot\vec{A}) - \vec{\nabla}^2\vec{A} \tag{16.24}$$

16.8.4 Solution to exercise 16.7.4

We simply check it by brute force.

$$\vec{\nabla}\times\vec{E} = -\vec{\nabla}\times\vec{\nabla}\phi \tag{16.25}$$

| recall that $\phi(\vec{r}) = \dfrac{1}{4\pi\epsilon_0} \displaystyle\int \dfrac{\rho(\vec{r}')}{|\vec{r}-\vec{r}'|} d^3\vec{r}'$

$$
\begin{aligned}
&\quad|\ \text{so need } \vec{\nabla}\frac{1}{|\vec{r}-\vec{r}'|} \text{ and use } |\vec{r}-\vec{r}'|=\sqrt{(x-x')^2+(y-y')^2+(z-z')^2}\\
&\quad|\ \text{then } \vec{\nabla}\frac{1}{|\vec{r}-\vec{r}'|}=\hat{x}\frac{\partial}{\partial x}+\cdots=-\frac{(x-x')\hat{x}+\cdots}{((x-x')^2+\cdots)^{3/2}}=-\frac{\vec{r}-\vec{r}'}{|\vec{r}-\vec{r}'|^3}\\
&=\vec{\nabla}\times\frac{1}{4\pi\epsilon_0}\int\rho(\vec{r}')\frac{\vec{r}-\vec{r}'}{|\vec{r}-\vec{r}'|^3}d^3\vec{r}' \quad &(16.26)\\
&\quad|\ \text{use } \vec{\nabla}\times(f\vec{A})=f(\vec{\nabla}\times\vec{A})-\vec{A}\times(\vec{\nabla}f)\\
&\quad|\ \text{where scalar } f=\frac{1}{|\vec{r}-\vec{r}'|^3} \text{ and vector } \vec{A}=\vec{r}-\vec{r}'\\
&=\frac{1}{4\pi\epsilon_0}\int\left[\frac{\rho(\vec{r}')}{|\vec{r}-\vec{r}'|^3}(\vec{\nabla}\times(\vec{r}-\vec{r}'))-\rho(\vec{r}')(\vec{r}-\vec{r}')\times\left(\vec{\nabla}\frac{1}{|\vec{r}-\vec{r}'|^3}\right)\right] \quad &(16.27)\\
&\quad|\ \text{So } \vec{\nabla}\times(\vec{r}-\vec{r}')=\vec{\nabla}\times\vec{r}=0 \text{ and } \vec{\nabla}\times\vec{r}'=0 \text{ because they are independent}\\
&\quad|\ \text{then } \vec{\nabla}\frac{1}{|\vec{r}-\vec{r}'|^3}\propto(\vec{r}-\vec{r}') \text{ then } (\vec{r}-\vec{r}')\times(\vec{r}-\vec{r}')=0\\
&=0 \quad &(16.28)
\end{aligned}
$$

16.8.5 Solution to exercise 16.7.5

1.
$$
\begin{aligned}
\vec{E}=-\vec{\nabla}\phi &= -\vec{\nabla}\frac{1}{4\pi\epsilon_0}\frac{\vec{p}\cdot\hat{r}}{r^2} \quad &(16.29)\\
&\quad|\ \text{with } \vec{p}=p\hat{z} \text{ then } \vec{p}\cdot\hat{r}=p\cos\theta\\
&= -\frac{p}{4\pi\epsilon_0}\vec{\nabla}\frac{\cos\theta}{r^2} \quad &(16.30)\\
&= -\frac{p}{4\pi\epsilon_0}\left[\hat{r}\frac{\partial}{\partial r}\left(\frac{\cos\theta}{r^2}\right)+\hat{\theta}\frac{1}{r}\frac{\partial}{\partial\theta}\left(\frac{\cos\theta}{r^2}\right)\right] \quad &(16.31)\\
&= -\frac{p}{4\pi\epsilon_0}\left[-\frac{2\cos\theta\hat{r}}{r^3}-\frac{\sin\theta\hat{\theta}}{r^3}\right] \quad &(16.32)\\
&= \frac{p}{4\pi\epsilon_0 r^3}\left(2\cos\theta\hat{r}+\sin\theta\hat{\theta}\right) \quad &(16.33)
\end{aligned}
$$

2. We add and subtract $\cos\theta\hat{r}$
$$
\begin{aligned}
\vec{E} &= \frac{p}{4\pi\epsilon_0 r^3}\left(3\cos\theta\hat{r}-\cos\theta\hat{r}+\sin\theta\hat{\theta}\right) \quad &(16.34)\\
&\quad|\ \text{recall that } p\cos\theta=\vec{p}\cdot\hat{r}\\
&\quad|\ \text{then by geometry, } -\cos\theta\hat{r}+\sin\theta\hat{\theta}=-\hat{z}\\
&\quad|\ \text{note that it is unit } \hat{z} \text{ because of Pythagoras theorem, } \sin^2\theta+\cos^2\theta=1\\
&= \frac{1}{4\pi\epsilon_0 r^3}\left(3(\vec{p}\cdot\hat{r})\hat{r}-p\hat{z}\right) \quad &(16.35)\\
&= \frac{1}{4\pi\epsilon_0 r^3}\left(3(\vec{p}\cdot\hat{r})\hat{r}-\vec{p}\right) \quad &(16.36)
\end{aligned}
$$

16.8.6 Solution to exercise 16.7.6

$$
\vec{\nabla}\cdot\vec{A} = \frac{\mu_0}{4\pi}\int\vec{\nabla}\cdot\frac{\vec{J}(\vec{r}')}{|\vec{r}-\vec{r}'|}d^3\vec{r}' \quad (16.37)
$$

Summary of Electromagnetism 157

| note that the divergence affects $\frac{1}{|\vec{r}-\vec{r}'|}$ only

$$= \frac{\mu_0}{4\pi} \int \vec{J}(\vec{r}') \cdot \vec{\nabla} \frac{1}{|\vec{r}-\vec{r}'|} d^3\vec{r}' \qquad (16.38)$$

| write $\vec{\nabla} \frac{1}{|\vec{r}-\vec{r}'|} = -\vec{\nabla}' \frac{1}{|\vec{r}-\vec{r}'|}$ then integrate by parts

$$= -\frac{\mu_0}{4\pi} \int \left[\vec{\nabla}' \cdot \left(\frac{\vec{J}(\vec{r}')}{|\vec{r}-\vec{r}'|} \right) - \frac{1}{|\vec{r}-\vec{r}'|} \left(\vec{\nabla}' \cdot \vec{J}(\vec{r}') \right) \right] d^3\vec{r}' \qquad (16.39)$$

| convert first term into surface integral by divergence theorem
| and the first term is zero as current is assumed to be localised
| the 2nd term has $\vec{\nabla}' \cdot \vec{J}(\vec{r}') = 0$ since current is stationary

$$= 0 \qquad (16.40)$$

Then now,

$$\vec{\nabla} \times \vec{B} = \vec{\nabla} \times (\vec{\nabla} \times \vec{A}) \qquad (16.41)$$
$$= \vec{\nabla}(\vec{\nabla} \cdot \vec{A}) - \vec{\nabla}^2 \vec{A} \qquad (16.42)$$

| but we have $\vec{\nabla} \cdot \vec{A} = 0$ from earlier

$$= -\frac{\mu_0}{4\pi} \int \vec{\nabla}^2 \frac{\vec{J}(\vec{r}')}{|\vec{r}-\vec{r}'|} d^3\vec{r}' \qquad (16.43)$$

$$= -\frac{\mu_0}{4\pi} \int \vec{J}(\vec{r}') \left(\vec{\nabla}^2 \frac{1}{|\vec{r}-\vec{r}'|} \right) d^3\vec{r}' \qquad (16.44)$$

| then $\vec{\nabla}^2 \frac{1}{|\vec{r}-\vec{r}'|} = -4\pi\delta^3(\vec{r}-\vec{r}')$

$$= \mu_0 \vec{J}(\vec{r}) \qquad (16.45)$$

16.8.7 Solution to exercise 16.7.7

Use Ampere's law: $\oint \vec{B} \cdot d\vec{l} = \mu_0 I$ to get $2Bl = \mu_0 Kl \implies B = \frac{\mu_0 K}{2}$. The direction is determined by the Right-Hand Rule. So for $z > 0$, $\vec{B} = -\frac{\mu_0 K}{2}\hat{y}$ and for $z < 0$, $\vec{B} = +\frac{\mu_0 K}{2}\hat{y}$.

16.8.8 Solution to exercise 16.7.8

1. Assume that the charges are spread out uniformly. We combine $C = \frac{Q}{V} = \frac{\epsilon_0 A}{w}$ and $A = \pi a^2$ to get $V = \frac{Qw}{\epsilon_0 \pi a^2}$. Then bring in $Q = It$ and $E = \frac{V}{w}$ to get $\vec{E} = \frac{It}{\epsilon_0 \pi a^2}\hat{z}$.

   Choose a circular Amperian loop in the $+\hat{\phi}$ direction with radius s,

$$\oint \vec{B} \cdot d\vec{l} = \mu_0 I + \mu_0 \epsilon_0 \int \frac{\partial \vec{E}}{\partial t} \cdot d\vec{A} \qquad (16.46)$$

$$B(2\pi s) = \mu_0 0 + \mu_0 \epsilon_0 \frac{I}{\epsilon_0 \pi a^2}(\pi s^2) \qquad (16.47)$$

$$B = \frac{\mu_0 I s}{2\pi a^2} \qquad (16.48)$$

| determine the direction by Right Hand Rule

$$\vec{B} = \frac{\mu_0 I s}{2\pi a^2}\hat{\phi} \qquad (16.49)$$

2.
$$\vec{S} = \frac{1}{\mu_0}\vec{E}\times\vec{B} = \frac{It}{\epsilon_0\pi a^2}\frac{Is}{2\pi a^2}(\hat{z}\times\hat{\phi}) \qquad (16.50)$$

| where $\hat{z}\times\hat{\phi} = -\hat{s}$

$$= \frac{I^2 ts}{2\pi^2\epsilon_0 a^4}(-\hat{s}) \qquad (16.51)$$

3.
$$\text{Power} = \vec{S}\cdot\vec{A} \qquad (16.52)$$

| set $s = a$

$$= \frac{I^2 ta}{2\pi^2\epsilon_0 a^4} 2\pi a w \qquad (16.53)$$

$$= \frac{I^2 wt}{\pi\epsilon_0 a^2} \qquad (16.54)$$

16.8.9 Solution to exercise 16.7.9

We have $\vec{E} = \frac{A}{r}\sin\theta\cos\left(\omega\left(t - \frac{r}{c}\right)\right)\hat{\theta}$ and $\vec{\nabla}^2\vec{E} = \mu_0\epsilon_0\frac{\partial^2\vec{E}}{\partial t^2}$.

1.
$$\text{RHS} = \mu_0\epsilon_0\frac{\partial^2\vec{E}}{\partial t^2} = -\frac{\omega^2}{c^2}\vec{E} \qquad (16.55)$$

$$\text{LHS} = \vec{\nabla}^2\vec{E} \qquad (16.56)$$
$$= \vec{\nabla}(\vec{\nabla}\cdot\vec{E}) - \vec{\nabla}\times(\vec{\nabla}\times\vec{E}) \qquad (16.57)$$

| note that $\vec{\nabla}\cdot\vec{E}\propto\frac{1}{r^2}$ so $\vec{\nabla}\cdot\vec{E}\approx 0$ in far field

$$\approx -\vec{\nabla}\times(\vec{\nabla}\times\vec{E}) \qquad (16.58)$$

| then $\vec{\nabla}\times\vec{E} = \frac{1}{r}\left(\frac{\partial}{\partial r}(eE_\theta)\right)\hat{\phi} = \frac{A}{r}\frac{\omega}{c}\sin\theta\sin\left(\omega\left(t - \frac{r}{c}\right)\right)\hat{\phi}$

| then $\vec{\nabla}\times(\vec{\nabla}\times\vec{E}) \approx -\frac{1}{r}\frac{\partial}{\partial r}\left(A\sin\theta\frac{\omega}{c}\sin\left(\omega\left(t - \frac{r}{c}\right)\right)\right)\hat{\theta}$

| $= \frac{\omega^2}{c^2}\frac{A}{r}\sin\theta\cos\left(\omega\left(t - \frac{r}{c}\right)\right)\hat{\theta}$

$$\approx -\frac{\omega^2}{c^2}\vec{E} \qquad (16.59)$$

Hence the wave equation is only satisfied in the far field region (see the derivation leading to equation (4.40)).

2. Take a plane wave (in complex representation) $\vec{E} = \vec{E}_0 e^{i(\vec{k}\cdot\vec{r}-\omega t)}$ with $\vec{E}_0 = E_0\hat{n}$ where $\hat{n}$ is the polarisation unit vector. For transverse wave $\hat{n}\cdot\vec{k} = 0$.

Then $\vec{\nabla}\cdot\vec{E} = (\vec{\nabla}\cdot\vec{E}_0)e^{i(\vec{k}\cdot\vec{r}-\omega t)} + \vec{E}_0\cdot(\vec{\nabla}e^{i(\vec{k}\cdot\vec{r}-\omega t)})$ and $\vec{\nabla}\cdot\vec{E}_0 = 0$ because $\vec{E}_0$ is a constant vector. So $\vec{\nabla}\cdot\vec{E} = \vec{E}_0\cdot(\vec{\nabla}e^{i(\vec{k}\cdot\vec{r}-\omega t)}) = \vec{E}\cdot i\vec{k} = 0$ since $\hat{n}\cdot\vec{k} = 0$.

Bibliography

[1] Berthold-Georg Englert. *Lectures On Quantum Mechanics - Volume 3: Perturbed Evolution.* World Scientific, 2024.

[2] Herbert Goldstein, Charles Poole, John Safko, and Stephen R Addison. *Classical mechanics.* American Association of Physics Teachers, 2002.

[3] Walter Greiner and S Soff. *Classical Electrodynamics.* Classical Theoretical Physics. Springer, New York, NY, 1998 edition, October 1998.

[4] David J Griffiths. *Introduction to electrodynamics.* Pearson, Upper Saddle River, NJ, 3 edition, December 1998.

[5] John David Jackson. *Classical electrodynamics.* John Wiley & Sons, 2021.

[6] Harald JW Muller-Kirsten. *Electrodynamics.* World Scientific Publishing Company, 2011.

[7] Wolfgang KH Panofsky and Melba Phillips. *Classical electricity and magnetism.* Courier Corporation, 2012.

Index

Note: *Italic* page numbers refer to figures and page numbers followed by 'n' denotes notes.

A
Ampere–Maxwell equation, 10
Ampere–Maxwell law, 4, 6, 132, 152
Ampere's law, 74, 150, 157
Angular momentum, 8
Arbitrary distribution, 51
    radiation from, *36*, 36–38

B
Basic scattering models, 44–48
    scattering by bound charge, 46–48
    scattering by free charge, 44–46
    Thomson scattering, 44–46
Biot-Savart's law, 5, 29, 50, 120, 148
Brewster's angle, 152

C
Canonical energy-momentum tensor, 138
Cartesian coordinates, 143
Cauchy's residue theorem, 13
Charge continuity equation, 3, 34, 81
Charge moving at constant velocity, 50
Classical atom, decay of, 51
Classical Field Theory, 99
Classical relativistic $U(1)$ gauge theory, 99
    Lagrangian description of, 99–111
Compton scattering, 46, 52, 125
Conservation laws, 3–8, 81–82
    angular momentum, 8
    charge continuity equation, 3
    energy continuity equation, 4–5
    momentum continuity equation, 5–7
Conservation of charge, 3
Conservation of energy, 4
Conservation of momentum, 6
Coulomb gauge, 11–12, 49, 116
Coulomb's law, 5, 27, 29, 146
    time-dependent generalisation of, 19

D
D'Alembert's operator, 67–68
Damped integral, 14n10
Damping force, 127
Derivative 4-vectors, 67–68
Dielectric sphere scatterer, 52
Dipole radiation, 126n2, 127
Dipole retarded potentials, satisfy Lorenz gauge, 51
Dirac delta function, 145
Dual field strength tensor, 83

E
Einstein relation, 65–67, 100
Electric dipole, 153
Electromagnetic waves, 151–152, 154
Electromagnetic (EM) waves scattering, 39–48
    basic scattering models, 44–48
    formalism, optical theorem, *40*, 40–44
    scattering problem, statement, 39–40
Electromagnetism, 146–158
    electric fields, matter, 147–148
    electrostatics, 146–147
    magnetic fields, matter, 149–150
    magnetostatics, 148–149
Electrostatics, 146–147, 153
Energy continuity equation, 4–5, 81
Equations of electrodynamics, 10
Equations of motion, 101, 102, 106
Euler-Lagrange equations, 101, 104, 106, 107

F
Faraday's law, 6, 9, 150, 152
Field momentum density, 7
Field strength tensor, 71
Formalism, optical theorem, *40*, 40–44
Fourier form, 13

G

Gauge-fixing function, 10
Gauge invariance, 96
Gauge transformations, 10–12, 139
    Coulomb gauge, 11–12
    Lorenz gauge, 12
Gauss law, 6, 9, 79, 146
Glashow-Weinberg-Salam (GSW) electroweak theory, 111
Green's function, 12, 13, 84

H

Hertzian dipole, 36, 51
    radiation from, *32*, 32–36
    retarded potentials, 51

I

Internal symmetry transformation, 11n3
Invariant hyperbola, *130*
Invariant interval, 61–63

J

Jefimenko's equations, 19–20, 119
    quasistatic approximation of, 50
Joule heating law, 150

K

K^0, meaning, 95
Klein–Nishina formula, 46, 52, 125
Kronecker delta, 155

L

Lagrangian density, 104, 105, 112
Lagrangian description, 99–111
    of EM field, 104–111
Lagrangian mechanics, 99–100
Laplace's equation, 147
Larmor's formula, 32, 36, 45, 88–93, 122
Legendre transform, 100–102
Length contraction, 59–60
Lienard–Wiechert potentials, 21–24, 27, 84–86
    in covariant form, 96
Line integral, 152
Lorentz covariant form, 71–93
Lorentz force, 72
Lorentz force law, 79
Lorenz gauge, 12, 49–51, 81, 117, 123

Lorentz invariants, 96
    in electrodynamics, 83
    equation, 78
    form, 84
Lorentz transformations, 61, 65, 68, 70, 75, 83, 95, 133, 134
Lorentz transformations, $\vec{E}$ and $\vec{B}$ fields, 73–76
    derivation of, 55–58
    of momentum 4-vector, *66*

M

Magnetic dipole radiation, 51
Magnetic field, surface current, 153
Magnetism, relativistic phenomena, 71–72
Magnetostatics, 148–149, 153
Maxwell's equations, 76–81, 95, 106, 132, 151
Maxwell stress tensor, 6, 7, 49
Mechanical energy density, 5
Mechanical momentum density, 7
Minkowski diagram exercises, 94
    invariant hyperbola, 94
    pole and barn paradox, 94–95
    relative simultaneity, 94
Minkowski force, 69, 70, 79, 82
Minkowski space, 71, 103
Minkowski spacetime diagram, 61, *62*, *63*
Momentum 4-vector, 65
Momentum continuity equation, 5–7, 81

N

Newton's first law, 69
Newton's second law, 69
Newton's third law, 6
Noether's current density, 108
Noether's theorem, 99–100, 106, 108
(Non-relativistic) equation of motion, 45

O

Ohm's law, 150
Optical theorem, *40*, 40–44, 52, 126

P

Photon, 95
Poincare group, 94, 129
Point charge, 21–29, 51
    in constant velocity, 26–29
    fields, 24–26
        of charged particle, constant velocity, 86–88

Index

Lienard's generalisation, Larmor's formula, 88–93
Lienard–Wiechert potentials, 84–86
radiation from, 30–32
retarded potentials, 21–24, *22*
Poisson equation, 9, 11, 116, 146
time-dependent generalisation of, 9
Potential formulation, electrodynamics, 9–20
gauge transformations, 10–12
Jefimenko's equations, 19–20
solutions, retarded potentials, 12–18
Poynting's theorem, 4, 5
Poynting vector, 5, 6, 30, 35, 38, 42, 153

Q

Quantum chromodynamics (QCD), 111
Quantum field theory (QFT), 111
Quantum optics, 111

R

Radiation, 30–38
acceleration parallel to velocity, 96
from arbitrary distribution, *36*, 36–38
from hertzian dipole, 32–36
from point charge, 30–32
Rayleigh cross section, 48
Relativistic mechanics, 69–70
Relativistic particle Lagrangian, 100
charged particle, external EM field, 100–104
Retarded potentials, 21–24, *22*, 49
satisfies Lorenz gauge, 50

S

Scattered power, 42
Scattering amplitude, 44
Special relativity, 55–70
basic consequences of, 59–61
Einstein's axioms for, 55
invariant interval, 61–63
length contraction, 59–60
Lorentz transformation, 55–58
Minkowski spacetime diagram, 61
tensor calculus, transition, 63–70
derivative 4-vectors and d'Alembert's operator, 67–68
Minkowski metric, 4-vectors, 63–65
momentum 4-vector and Einstein relation, 65–67
relativistic mechanics, 69–70
time dilation, 59
velocity addition, 60–61
Spherical coordinates, 143–144

T

Taylor expansion, 50
Thomson scattering, 44–46, 52, 125
Time dilation, 59

V

Vector calculus, 143–145
3 coordinate systems, 143–145
Cartesian coordinates, 143
cylindrical coordinates, 144
spherical coordinates, 143–144
vector identities, 144–145
Vector identities, 144–145
Dirac delta function, 145
integral theorems, 145
product rules, 144–145
second derivatives, 145
triple products, 144
Velocity 4-vector, 65
Velocity addition, 60–61

W

Work-energy theorem, 4

Printed in the United States
by Baker & Taylor Publisher Services